CW00727557

Recent Advances in
Animal Nutrition – 1992

In the same series:

Recent Advances in Animal Nutrition - 1988
Edited By W. Haresign and D. J. A. Cole

Recent Advances in Animal Nutrition - 1989
Edited by W. Haresign and D. J. A. Cole

Recent Advances in Animal Nutrition - 1990
Edited by W. Haresign and D. J. A. Cole

Recent Advances in Animal Nutrition - 1991
Edited by W. Haresign and D. J. A. Cole

Related titles:

Avian Incubation
22nd Poultry Science Symposium
Edited by S. G. Tullet

Biotechnological Innovations in Animal Productivity
BIOTOL

Biotechnological Innovations in Food Processing
BIOTOL

Biotechnology in Growth Regulation
Edited by R. B. Heap, C. G. Prosser,
G. E. Lamming

Control of Fat and Lean Deposition
51st Nottingham Easter School in Agricultural Sciences
Edited by P J Buttery, K N Boorman and
D B Lindsay

Feed of Non-Ruminant Livestock
Edited by J. Wiseman

Feedstuff Evaluation
50th Nottingham Easter School in Agricultural Sciences
Edited by J. Wiseman, D. J. A. Cole

Food Legislation of the UK, 2nd edition
D. J. Jukes

Food Legislative System in the UK
S. J. Fallows

Leaness in Domestic Birds
Edited by B. Leclercq and C. C. Whitehead

Manual of Australian Agriculture 5th edition
Edited by R. L. Reid

New Techniques in Cattle Production
Edited by C. J. C. Phillips

Non Traditional Feed Sources for Use in Swine Production
Edited by Phillip A. Thacker and R. N. Kirkwood

Nutrition and Lactation in the Dairy Cow
46th Nottingham Easter School in Agricultural Sciences
Edited by P. C. Garnsworthy

Pig Production in Australia
Edited by J. A. A. Gardiner, A. C. Dunkin, L. C. Lloyd

Recent Advances in Turkey Sciences
21st Poultry Science Symposium
Edited by C. Nixey and T. C. Grey

Recent Developments in Poultry Nutrition
Edited by W. Haresign and D. J. A. Cole

Structure and Function of Domestic Animals
W. Bruce Currie

Swine Nutrition
E. R. Miller, D. E. Ullrey and A. J. Lewis

Transgenic Animals
Edited by neal First and Florence P. Haseltine

Veterinary Vaccines
Edited by A. R. Peters

Voluntary Food Intake of Farm Animals
J. M. Forbes

If you wish to order any of the above titles please contact your local bookseller, or in case of difficulty:

Reed Book Services Ltd
PO Box 5
Rushden
Northants
NN10 9YX
Tel: (0933) 410511
Fax: (0933) 50284

Recent Advances in Animal Nutrition

1992

P.C. Garnsworthy, PhD
W. Haresign, PhD
D.J.A. Cole, PhD
University of Nottingham School of Agriculture

Butterworth-Heinemann Ltd
Linacre House, Jordan Hill, Oxford OX2 8DP

 PART OF REED INTERNATIONAL BOOKS

OXFORD LONDON BOSTON
MUNICH NEW DELHI SINGAPORE SYDNEY
TOKYO TORONTO WELLINGTON

First published 1992

British Library Cataloguing in Publication Data
University of Nottingham Feed Manufacturers
Conference (26th 1992 Nottingham)
Recent advances in animal nutrition – 1992
– (Recent advances in animal nutrition)
I. Title II. Garnsworthy, P.C. III. Haresign, W.
IV. Cole, D.J.A. V. Series
636.08

ISBN 0 7506 0714 9

Printed and bound in Great Britain by Redwood Press, Melksham, Wiltshire

PREFACE

The animal feed compounding industry is facing major challenges with overcapacity in milling and the downturn in economic returns from animal production. At the same time consumers are becoming more demanding about product quality and are increasingly critical of the possible environmental impact of animal production systems. The various chapters within the proceedings of the 26th University of Nottingham Feed Manufacturers were designed to address such issues.

A number of plant protein sources contain anti-nutritional factors which impair nutrient utilisation and hence performance in non-ruminant animals. A knowledge of such factors is essential for determining the extent to which home-produced plant proteins can be used to replace more traditional imported protein sources. Hockburn results in the downgrading of broilers; it is a problem with a number of contributory causes and one chapter specifically addresses the extent to which it can be influenced by dietary composition. A further chapter re-evaluates the relationships between protein quality and amino acid utilisation in poultry.

The extent to which nutrient requirements are influenced by the genetic merit of the animal are now well established in non-ruminant species. However, recent evidence suggests that it is also necessary to alter nutritional inputs to dairy cows in accordance with their genetic merit. While the move to describing protein requirements of ruminants in terms of degradable and undegradable fractions recognised interactions between energy and nitrogen supplies within the rumen and was a major step forward in diet formulation, this particular scheme fails to account for some of the responses of dairy cows to protein with diets high in soluble carbohydrates. The recently introduced metabolisable protein system represents a major advance in overcoming these limitations and improving the precision of diet formulation for ruminants.

The excretion of large quantities of nitrogen and phosphorus in the faeces of farm animals now represents a major pollution threat to the environment in a number of EEC countries. In the past diets have been designed on a least-cost system rather than one which maximises nutrient utilisation and minimises pollution. However, the pressure to change is increasing and one chapter details the way in which this might be achieved. Although the EEC has banned the use of steroid growth promoters in animal production, a number of feed additives which are known to have growth promoting properties are now coming under increasing scrutiny. Evidence was presented on the impact that these materials have on nutrient utilisation and waste production to support their continued use, not only to enhance animal performance but more importantly for reducing the environmental pollution.

EEC legislation is demanding that feed manufacturers supply compound feeds with tight tolerances for energy and protein content. Many feed manufacturers have erroneously assumed that increasing the number of raw materials in a finished feed would reduce formulation errors. New computing packages now make it possible for compounders to predict variation in diet composition and to take the necessary steps to produce diets within the permitted legal tolerances.

It is now accepted within the EEC that there is urgent need to reform the common agricultural policy (CAP). In future, it is also likely that world factors will have an increasing role on economic matters through the general agreement on tariffs and trade (GATT) discussions. While the nature of these influences has yet to be finalised, one chapter attempts to predict the possible impact of such reforms on the UK animal feed compounding industry.

Recent years have seen major changes in the design of machinery which have changed the temperature/time relationships at various stages of the feed manufacturing process. Such changes can have a significant impact on the stability of heat-labile nutrients and the options available to the compounder to compensate for these effects. The final chapter provides an up-to-date account of the array of changes in legislation (domestic and EEC) which affect the feed compounding industry.

All chapters are written in a clear and informative manner and should be of considerable benefit to research workers, livestock advisors and anyone involved in the animal feed industry.

The organisers of the Conference and the University of Nottingham are grateful to BP Nutrition for the continued support which they provided towards the running of this Conference. Thanks are also due to John Burgess for the skill with which he did the typesetting of these proceedings.

P C Garnsworthy
W Haresign
D J A Cole

CONTENTS

I

Non-ruminant Nutrition

1

ANTINUTRITIONAL FACTORS IN THE PLANT PROTEINS OF DIETS FOR NON-RUMINANTS

J. HUISMAN and G.H. TOLMAN

TNO-Institute for Animal Nutrition and Physiology (ILOB), P.O. Box 15, 6700 AA Wageningen, The Netherlands

Introduction

The major sources of protein in diets for monogastric animals are vegetable. The main one is soya imported from outside the EC. Due to the growing interest in Europe in becoming more self-supporting in protein supply for animals, plant protein sources other than soya, such as pea, bean and rapeseed protein, are used in diets for monogastric animals. The nutritional value of plant protein sources depends, in addition to the chemical composition, on the extent to which nutrients are digested, absorbed and utilized. Various factors may interfere with digestibility, absorption and utilization of nutrients. Two major classes of factors can be distinquished in depressing digestibility: a lack of appropriate enzymes present in the gastro-intestinal tract and the presence of substances hampering digestion, absorption or utilization of nutrients. The latter class of factors is called antinutritional factors (ANFs). This paper deals with the interference by ANFs.

ANFs can be classified as non-fibrous natural substances having negative effects on growth or health of man and animals. In this definition fibre is excluded because in human food fibre may be classified as a positive health factor. Further, fibre may have some energy value when digested in the large intestine. Mycotoxins, because they are classified as contaminants and factors originating from processing, are excluded (Yannai, 1980).

Most ANFs give the plant a natural protection against attacks of moulds, bacteria, insects and birds (Bond and Smith, 1989; Broadway *et al.*, 1986; Etzler, 1986; Jaffé, 1980; Janzen *et al.*, 1976; Ryan, 1983; Liener and Kakade, 1980; Pistole, 1981). These reports show that the defensive effect of ANFs in plants and seeds seems to be related to disturbances in digestive processes in micro-organisms and insects. Because there are similarities in digestive processes in farm animals, micro-organisms and insects, ANFs can be expected to disturb the digestive processes in farm animals in a similar way to micro-organisms and insects.

ANFs can be classified in different ways. In the following scheme they are classified on the basis of their effects on the nutritional value of feedstuffs and on the biological response in the animal:

- - factors with a depressive effect on protein digestion and on the utilization of protein (trypsin and chymotrypsin inhibitors, lectins, phenolic compounds, saponins);
- - factors with a negative effect on the digestion of carbohydrates (amylase inhibitors, phenolic compounds, flatulence factors);
- - factors with a negative effect on the utilization of minerals (glucosinolates, phytic acid, oxalic acid, gossypol);
- - factors that inactivate vitamins or cause an increase in the animals' vitamin requirement;
- - factors that stimulate the immune system (antigenic proteins).

The role of ANFs in animal nutrition may become more important in the future. This is related to the general expectation that farm animals of tomorrow will grow faster and deposit more protein in the body, because of advances in animal breeding, health care and housing (Webb, 1989). These animals will have a higher requirement of digestible protein.

Because in modern pig breeds the feed intake capacity has not been increased yet (Brandt *et al.*, 1985; Webb and Curran, 1986), feedstuffs with a high (and highly digestible) protein content will become more important. As most plant protein sources contain ANFs it is necessary to study different aspects of ANFs such as the mode of action in the animal, the maximum tolerance levels in the diets, differences among animal species and improvement and standardization of analytical methods to detect ANFs. This information is the basis for the development of methods to reduce the negative effects. Various aspects of these points will be discussed in this chapter.

Occurrence of ANFs in seeds

Recent reviews, reports and books on the presence and distribution of ANFs in various seeds have been presented by Cheeke and Shull (1985); Chubb (1982); Friedman (1986); Huisman (1989); Huisman *et al.* (1989); Huisman and Jansman (1991); Liener (1980, 1989); Pusztai (1989), Rackis *et al.* (1986); Savage and Deo (1989); Sissons and Tolman (1991). The various seeds contain several different ANFs (Chubb, 1982; Huisman, 1989; Huisman *et al.*, 1990a; Liener, 1981; Savage and Deo, 1989). When these seeds are fed to animals the negative effects may be attributed to a combination of the effects of different ANFs. In Table 1.1 the ANFs in various seeds are listed. It shows that trypsin and chymotrypsin inhibitors and lectins are most important in the legume seeds soya, peas and beans, but some varieties of rye and triticale may also contain moderate levels of trypsin inhibitors. Tannins are mainly present in sorghum, in some barley varieties, in the coloured flowering varieties of beans and peas and in rapeseed.

Glucosinolates and sinapins are important in rapeseed. Alkaloids are present in lupins, whereas soya may be contaminated with alkaloids originating from Datura. Gossypol is present in cotton seed. Antigenicity of proteins is found in soya, peas and beans.

Table 1.1. ANTINUTRITIONAL FACTORS IN CEREALS AND SEEDS

Cereals/seeds	*ANFs*			
	trypsin inhibitors	*lectins*	*polyphenolic compounds*	*other ANFs*
Cereal grains				
Wheat, Rice, Maize	-/+	-	-	-
Rye	-/+/++	-	-	-
Triticale	-/+/++	-	-	-
Barley	-/+	-	-/+/++	-
Sorghum	-/+	-	+/++/+++	-
Legume seeds				
Soya	++/+++	++	-	++/+++[A,C]
Vicia faba bean	+	+	+/++/+++	+/++/+++[B]
Ph. vulgaris bean	-/+/++	+/++/+++	+/++	+/++/+++[A]
Pisum sativum	+/++	+/++	+/++	-
Lentils, cowpeas, chick peas	+/++	+/++	-/+/++	-
Lupins	-	-	-	+/++/+++[C]
Other seeds				
Rapeseed	-	-	+/++	+/++/+++[D]
Sunflower seed	-/+	-	+/++[E]	-
Cotton seed	-/+	-	-	+/++/+++[F]
Peanut	-	-	+/++[G]	-

- below detection level; + low level; ++ medium level; +++ high level.
Different varieties of the same material may have different characteristics.
A, antigenic proteins; B, vicine/convicine; C, alkaloids; D, glucosinolates and sinapins; E, phenolic compounds (3-3.5%); F, gossypol; G, 16-18% in the shell around the nut.

Major effects of ANFs in monogastric animals

The major effects of ANFs on physiological processes in monogastric animals are given in Table 1.2

TRYPSIN INHIBITORS

General

Research into protease inhibitors started more than fifty years ago with the discovery by Read and Haas (1938) that aqueous extracts of defatted soya bean flour inhibited the ability of trypsin to break down gelatin. Subsequently Kunitz (1945, 1946) isolated the Kunitz soya bean trypsin inhibitor family (STI) and Bowman (1944) and Birk (1961) the Bowman-Birk trypsin inhibitor family (BBI). STI primarily inhibits trypsin and weakly inhibits chymotrypsin (Birk, 1987, 1989; Liener and Kakade, 1980). BBI inhibits both trypsin and chymotrypsin (Birk, 1987, 1989; Liener and Kakade, 1980).

Trypsin inhibitors (TIs) are widely distributed in plant seeds (Liener and Kakade, 1980). Most legume seeds contain high levels of these inhibitors (Table 1.1). The main TI in raw soya beans is STI, whereas BBI is present at a lower level. In other legumes mainly BBI and BBI-like forms are present. The levels of TIs and chymotrypsin inhibitors in raw beans and peas are lower than in raw soya beans (Griffiths, 1979; Newton and Hill, 1983). Huisman and Jansman (1991) have summarized various reports and found that, relative to raw soya beans, trypsin inhibitor activity (TIA) in raw peas was about 12% and in common beans about 38%.

Mode of action in the animal

STI and BBI are peptides which can form stable, inactive complexes with the proteolytic pancreatic enzymes trypsin and chymotrypsin (Birk, 1989; Liener and Kakade, 1980; Gallaher and Schneeman, 1986). Due to this complex formation, the activity of the trypsin and chymotrypsin decreases (Liener and Kakade, 1980; Rackis and Gumbmann, 1981; Rackis *et al.*, 1985). Inactivation of trypsin in the gut induces endocrine cells in the intestinal mucosa to release more of the hormone cholecystokinin-pancreozymin (CCK-PZ) which stimulates the pancreas to produce more digestive enzymes such as (chymo)trypsin, amylase and elastase (Birk, 1989; Liener and Kakade, 1980). Due to the enhanced enzyme production, the relative pancreas weight increases in small animal species due to hypertrophia and hyperplasia (Birk, 1989; Gallaher and Schneeman, 1986; Gumbmann *et al.*, 1985; Liener and Kakade, 1980). Pancreatic hypertrophy and hyperplasia are not observed in larger animal species such as growing pigs, dogs and calves (reviewed by Liener and Kakade, 1980 and Gallaher and Schneeman, 1986). Hypertrophy of the pancreas is not only found in relation to soya TI but also in rats and chicks fed on *Vicia faba* or purified TI from faba beans (Kardivel and Clandinin, 1974; Abbey *et al.*, 1979) and in chicks fed *Pisum sativum* (Johns, 1987; Huisman *et al.*, 1989). The negative feedback mechanism regulating the secretion of pancreatic enzymes found in rats also exists in pigs and calves, but without causing pancreatic hypertrophy (reviewed by Gallaher and Schneeman, 1986). Liener and Kakade (1980) related the relative weight of the pancreas of various species to the occurrence of pancreatic hypertrophy due to soya TI. They observed that those animal species in which the relative pancreatic weight exceeded 0.3%, became hypertrophic. Those animals whose relative pancreas weight was below this value did not respond in terms of pancreatic hypertrophy.

It is not clear whether CCK-PZ is the only factor causing hypertrophy of the pancreas, because Struthers *et al.* (1983) found in rats after infusing CCK in the blood circulation, lesser hypertrophy than after feeding raw soya bean. Naim *et al.* (1982) demonstrated that TI may not be the only factor in soyabeans causing hypertrophy in

Table 1.2. MAJOR EFFECTS OF ANFs IN MONOGASTRIC FARM ANIMALS

Antinutritional factor	*Major* in vivo *effect.*
Proteins	
Trypsin inhibitors	reduction of activity of (chymo)trypsin, pancreas hypertrophy, increased secretion of pancreatic enzymes.
Lectins	gut wall damage, immune response, increased loss of endogenous protein.
Amylase inhibitors	interference with starch digestion.
Antigenic proteins	interference with the gut wall integrity, immune response.
Polyphenols	
Tannins	formation of protein-carbohydrate complexes, interference with protein and carbohydrate digestibility.
Glycosides	
Vicine/convicine	haemolytic anaemia, interference with the fertility and hatchability of eggs.
Saponins	haemolysis, effects on intestinal permeability.
Glucosides	
Glucosinolates	impaired iodine utilization, affected thyroid and liver, reduced palatability and growth.
Alkaloids	
Quinolizidine (lupin alkaloids) Scopolamine and hyoscyamine (alkaloids from Datura)	neural disturbances, reduced palatability.
Other ANFs	
Phytate	forms complexes with minerals and protein, depresses absorption of minerals
Gossypol	anaemia due to Fe-complexation, reduced egg weight.
Sinapins	fishy odour in eggs (taint).
Flatulence factors	gastro-intestinal discomfort.

rats. Green *et al.* (1986) found that when amino acid availability is inadequate, pancreatic growth and adaptive response to CCK is impaired and protease secretion is insufficient to normalize CCK secretion, resulting in sustained elevation of the plasma CCK level. Indications that pancreatic hypertrophic effects and enzyme secretion are also influenced by protein quality and levels of protein in the diets can be found in studies by Green *et al.* (1986); Green and Nasset (1983) and Hasdai *et al.* (1989).

It has been shown in various studies that feeding raw soya bean meal to chickens causes a reduced weight gain (Liener and Kakade, 1980). The growth reduction cannot be fully attributed to the STI and BBI in the raw soya bean meal. The most logical explanation for growth reduction would be that the intestinal proteolytic enzymes trypsin and chymotrypsin are inactivated by the STI and BBI, thus interfering with the digestive processes in the intestinal tract. However, Liener *et al.* (1949) demonstrated that addition of STI to a pre-digested diet still caused growth depression. This shows that inactivation of the intestinal enzymes is not the only cause of growth depression. Also, the addition of STI to the same level as in raw soya bean meal did not cause the same growth reduction (Liener *et al.*, 1949; Gertler *et al.*, 1967; Nitsan and Gertler, 1972). Moreover, Kakade *et al.* (1973) found that 40% of the growth depression could be contributed to the STI. The factors responsible for the remaining 60% growth depression still have not been defined.

Various reports suggest that the negative effects of STI and BBI are related to the loss of endogenous protein due to a stimulated secretion of pancreatic enzymes (Gallaher and Schneeman, 1986; Liener and Kakade, 1980). The pancreatic enzymes trypsin and chymotrypsin are rich in cystine. Methionine can act as a methyl donor for cystine and the excessive synthesis of cystine may lead to a situation in which less methionine is available.

Various studies have demonstrated that, together with growth inhibition, due to feeding raw soya bean meal or trypsin inhibitors, pancreatic hypertrophy occurs (Birk, 1989; Liener and Kakade, 1980; Liener *et al.*, 1985; Nitsan, 1991; Nitsan and Alumot, 1965). However, these conclusions may be re-considered after examination of the results in Table 1.3.

Table 1.3. PANCREATIC HYPERTROPHY IN RATS FED SOYA BEAN MEAL

Dietary protein	*Final body weight* (g)	*Pancreas weight*	
		relative (%) *of body weight*	*absolute* (g)
Casein	147	0.56	0.83
Raw soya bean meal	89	0.85	0.76
Heated soya bean meal	148	0.50	0.74

Adapted with modifications from Green et al., 1986

When only the relative figures are used, the conclusion should be that there is a pancreatic hypertrophy due to feeding raw soya bean meal. However, when the absolute figures were used no pancreatic hypertrophy was observed. In this example the hypertrophic effect is related to reduced body weight, but not to changes in absolute pancreas weight. Therefore, the relevance of relative enlargement of the pancreas may be questioned. The effects of feeding raw soya bean meal to laying hens are not clear. Saxena *et al.* (1963) and Summers *et al.* (1966) found no negative effects on egg production. However, Rogler and Carrick (1964) and Nesheim and Garlich (1966) found a lower egg production with raw soya bean meal compared with feeding a heated soya bean meal and fish meal diet.

Most of the studies on the effects of TIs on weight gain, protein efficiency ratio (PER), digestibility and pancreas have been carried out with small animal species such as rats, mice and chickens. Knowledge on the way TI acts in animals is mainly based on data from these small animals and not on data from larger monogastric farm animals such as pigs. However, it is pertinent to note that striking results have been obtained in pigs, calves and guinea pigs when poorly digestible protein sources containing high levels of TIs were fed. Hasdai *et al.* (1989) found no effect on absolute pancreas weight in guinea pigs, but a reduced secretion of pancreatic enzymes when raw soya flour was fed instead of heated soya flour. Huisman *et al.* (1990b) found a reduced weight of the pancreas in piglets but hypertrophy of the pancreas in rats when raw Phaseolus beans were fed instead of heated beans. Further, in piglets the protein digestibility of raw beans was much lower than in rats. Le Guen *et al.* (1991) found a reduced enzyme activity in the pancreas of piglets fed raw peas with relatively high ANF contents and a low protein digestibility as compared with a pea protein isolate with very low levels of ANFs and a higher protein digestibility. Khorasani *et al.* (1989) recorded a reduced secretion of pancreatic enzymes in calves when soya protein isolate was fed instead of milk protein, but the protein digestibility of the soya protein diet was lower than for the milk protein diets.

These results indicate that in pigs, calves and guinea pigs the stimulating effect of TIs on the secretion of pancreatic enzymes is not present when poorly digestible proteins are fed. In this respect these animals react differently from small animals like rats and chickens.

Reduction of negative effects of trypsin inhibitors

In rats and chicks, it was demonstrated that the negative effect due to soya TI could be reduced significantly by including additional amino acids (methionine plus threonine, valine and lysine) in the diet (Borchers, 1961, 1962; Khayamhashi and Lyman, 1966). However, the possibility of partly abolishing ANF effects by supplementing with amino acids has only been demonstrated in rats and chickens, and not in pigs. It is questionable whether in pigs pancreatic secretion is affected to the same extent as in rats and chickens and whether the negative effect of soya TI can also be reduced by adding amino acids. For this, there is insufficient information about the response of the pancreas in pigs to TIs.

Until now, applying heat has been the most successful method of reducing the TIA. With heat the protein, including the proteinaceous ANFs (TIA, lectins and amylase inhibitors), will be degraded. Efficiency of the heat treatment depends of various factors, such as the temperature applied, duration of heating, the use of pressure during heating, particle size and moisture content (Liener and Kakade, 1980; Gupta, 1987; Rackis *et al.*, 1986; Van der Poel, 1989). Overheating of feeds may reduce the nutritional value due to the Maillard reaction. With chemical treatment, formation of lysinoalanine or methionine sulphoxide may occur (Yannai, 1980). Toasted soya flour usually contains about 15% of the TI activity present in raw soya bean meal (Rackis *et al.*, 1986). Heat-treated products will contain relatively more BBI than STI because STI is more heat-sensitive than BBI. This may have physiological consequences, because STI inhibits only trypsin while BBI inhibits trypsin as well as chymotrypsin.

LECTINS

Mode of action in the animal

Lectins are proteins which are mostly present in the form of glycoproteins (Jaffé, 1980). They are characterized by their unique capability to bind sugar components. The affinity to sugar components may differ among the various lectins. The structure and chemical composition of many lectins have been intensively studied (reviewed by Goldstein and Poretz, 1986). The primary effect of lectins is related to the fact that they bind to the mucosa of the intestinal wall. This binding can result in damage to the intestinal epithelial cells, which can result in a decreased absorption of nutrients, a change in the activity of brush border enzymes and hypersecretion of endogenous protein due to shedding of damaged cells, increased production of mucins and a loss of plasma proteins to the intestinal lumen (reviewed by Kik *et al.*, 1989; Jaffé, 1980; Pusztai, 1989). Moreover, there are indications that toxic lectins affecting the brush border are transported into the systemic circulation, which may cause production of specific lectin antibodies of the IgG class (reviewed by Huisman and Jansman, 1991 and Pusztai, 1989). All in all, these effects may cause a decreased nutrient digestibility, decreased nitrogen retention and sometimes scour. These effects lead to reduced weight gain and a less efficient feed conversion. There is a broad variation in toxicity among the lectins of the various seeds. Lectins from common beans are very toxic (Liener, 1986; Pusztai, 1989). Pea lectins were found to be non-toxic in piglets by Bertrand *et al.* (1988), but they were found to be toxic by M.J.L. Kik (personal communication) in piglets and by Jindal *et al.* (1982) in rats. M.J.L. Kik (unpublished results) compared the toxicity of various lectins in piglets and found that the lectins from Phaseolus beans were distinctly more toxic then those of peas and faba beans. Jaffé (1980) reports that soya bean lectins are less toxic than those of kidney bean. One example of a non-toxic lectin is the tomato lectin (Kilpatrick, 1985). These results demonstrate that for a nutritional evaluation it is important to study the toxicity of each type of lectin separately. Extrapolation of the results reported in the literature to practical circumstances for monogastric animals is seriously hampered for various reasons:

- - most research is carried out with diets containing high levels of lectins. These levels are generally higher than those present in practice. It is likely that the effects will be less pronounced when practical diets are used.
- - much research into the effects of lectins is being carried out with rats and other laboratory animals. It is not clear whether these results can be extrapolated to other species because there are reports showing distinct differences among animal species in sensitivity to lectins (Jaffé, 1980) or to other ANFs (Combs *et al.*, 1967; Huisman *et al,* 1990b,c; Jaffé, 1980; Van der Poel *et al.*, 1990a,b; Yen *et al.*, 1977).

Reduction of the negative effects of lectins

As with trypsin inhibitors, heat treatment is the method most frequently used to reduce lectin activities. The various possiblities of reducing lectin activity as well as conditions of treatments have been reviewed by Van der Poel (1990a,b). Table 1.4 shows some effects of treatment on inactivation of trypsin inhibitors and lectins in Phaseolus beans (also see Table 1.5).

Table 1.4. EFFECT OF TREATMENT ON REDUCTION OF TIA AND LECTIN CONTENT IN PHASEOLUS BEANS (EXPRESSED IN % OF ORIGINAL ACTIVITY/CONTENT)

Treatment	*Inactivation* (%)	
	TIA	*Lectins*
Steam treatment (100 °C; > 15 min)	65-97	90-100
Autoclaving (121 °C; > 15 min)	85-100	99-100
Dry roasting (various temp./time combinations)	54-82	85-99
Extrusion (145 °C; 16 s)	78-98	93-98

Adapted from Van der Poel (1990b).

The results show that lectins are more sensitive to heat than trypsin inhibitors. It is also clear that inactivation of TI and lectins depends on type of treatment. With dry roasting the inactivation was less effective than with steam heating, autoclaving and extrusion.

ANTIGENIC PROTEINS

Mode of action in the animal

Antigenic proteins in feed or raw materials are macromolecular proteins or glycoproteins capable of inducing a humoral immune response when fed to animals. In this case humoral immune response means the synthesis of specific polyclonal antibodies secreted in body fluids, such as blood, for eliminating the antigenic protein. The humoral immune response, together with the cellular immune response, forms the specific immune system, which is one of the defence mechanisms of man and animals to keep the integrity of the body. Other types of defence mechanisms are the physico-chemical barrier of the epithelium and the non-specific immune response (humoral factors like acute-phase proteins, complement and cytokinins, and cellular factors like phagocytes, granulocytes and natural killer cells). As a result of the specific immune response man and animal can become immune to the antigen. When the antigen is an infectious microbial agent (viruses, bacteria, fungi or parasites) immunity is a desired effect, because it combats the infectious agents so that the infection is of limited duration and leaves very little permanent damage (Roitt *et al.*, 1985). In the case of a feed antigen it is questionable whether an immune response is attractive. Contrary to a single infection by microbial antigens, man and animals are continuously exposed to feed antigens in their diet. This continuous exposure increases the chance that the immune response develops into an acute hypersensitivity reaction (the occurrence in an excessive or inappropriate form of an adaptive immune response, causing tissue damage) or into a chronic hypersensitivity reaction. The latter one is less extreme then the acute hypersensitivity reaction, but remains present as long as the antigens are fed. An acute hypersensitivity reaction such as the type I (IgE-mediated immediate hypersensitivity; reaction within 2 hours) or type IV (T cells-mediated delayed hypersensitivity; reaction after several hours or days) is rare, but chronic hypersensitivity seems to be very common when antigenic proteins are present in the diets.

Phenomena linked with chronic hypersensitivity are increased endogenous protein secretion (mucus, secretory immune globulin A (sIgA), epithelial cells), changes in gastro-intestinal motility, disturbances in gut architecture, changes in gut permeability and histological changes in gut tissue (reviewed for soya proteins in calf diets by Sissons and Tolman, 1991). The overall result is a lower apparent protein digestibility, sometimes scour and disappointing animal performance. Bush *et al.* (1991) showed that pea protein, as soya protein, can be responsible for immunologically mediated negative effects in preruminant calves. Le Guen *et al.* (1991b) showed that young piglets given a raw pea-based diet, developed circulatory antibodies against pea legumin and vicilin. However, piglets appear to be less sensitive to feed antigens than preruminant calves (Sissons, 1989).

The detection of specific antibodies in the serum of animals implies the uptake and presentation to T-cells of antigens (reviewed by Van Dijk *et al.*, 1988). Antigen uptake is possible when protein digestion is still incomplete and macromolecules stay intact. Due to non-specific absorption by M-cells in the ileum, defects in the mucosal barrier, IgA immunodefficiency or a high concentration of antigens in the gut lumen antigens can be absorbed. Summarizing, the negative effects of feed antigens might be:

- - a decreased utilization of feed proteins, because part of the protein is absorbed as intact macromolecular proteins (and hence presumably poorly or not utilizable) instead of amino acids and small peptides;
- - an increased maintenance requirement due to activation of the immune system;
- - an increased secretion of endogenous protein as the result of a chronic local hypersensitivity reaction, resulting in a decreased apparent protein digestibility. This is because part of the endogenous protein is digested and resorbed, but seems not to be utilized as efficiently as feed protein as a result of endogenous nitrogen losses in the faeces and urine;
- - the risk of inducing in some susceptible animals an acute hypersensitivity reaction, resulting in severe diarrhoea, poor performance or even death.

In view of these phenomena activation of the immune system by feed antigens does not seem likely. There are two ways to prevent stimulation of the immune system. The first is not to feed antigenic raw materials or to process the raw materials in such a way that the antigenicity of the raw material is destroyed. The second way is to induce oral tolerance by feeding antigens. This means that the animal develops a state of specific immunological non-responsiveness to the antigen. Induction of tolerance depends on factors like type of antigen, dose of the antigen fed to induce tolerance, age of the animal, nutritional status of the animal and the presence of intestinal infections (Mowat, 1987). Induction of tolerance has been described for most laboratory animals and piglets (Miller *et al.*, 1984; Miller *et al.*, 1991), while preruminant calves seem to fail to develop oral tolerance (Heppell *et al.*, 1989; Sissons and Tolman, 1991).

Inactivation of the antigenicity

Inactivation of antigenicity by heat treatment is not appropriate, because the process has to be so severe that protein quality is impaired (Anderson *et al.*, 1979). Table 1.5 summarizes the effect of toasting on ANFs in soya flour. It shows that trypsin inhibitors and lectins are very heat-sensitive, whereas antigenicity is much less affected.

Table 1.5. ANF CONTENTS IN UNTOASTED (SFu), SLIGHTLY TOASTED (SFl) AND NORMALLY TOASTED (SFn) SOYA FLOURS AS WELL AS IN ETHANOL/WATER-EXTRACTED SOYA-CONCENTRATE (SC); PDI = PROTEIN DISPERSIBILITY INDEX

Product	*PDI* (%)	*Trypsin inhibitor activity* (mg/g)[1]	*Lectins* (mg/g)[2]	*Antigens* (mg/g)[3]
SFu	90	23.9	7.3[4]	610
SFl	70	19.8	4.5	570
SFn	20	3.1	0.05	125
SC	6	2.5	<0.0001	<0.02

[1] Expressed as mg inhibited bovine trypsin per gram of product (Van Oort *et al.*, 1989).
[2] ELISA procedure.
[3] ELISA procedure by using anti-total soya protein IgG raised in calves by oral immunization with slightly toasted soya flour (PDI 70).
[4] A lectin content between 10 and 20 is more common for native soya protein.

More successful ways to inactivate antigenicity are chemical or enzymatic treatments. Sissons *et al.* (1982) showed that a hot aqueous ethanol extraction of soya flour produces a soya concentrate low in antigenicity. Most commercially available soya concentrates are now ethanol water-extracted soya concentrates. An example of ANF content of such a commercial soya concentrate is presented in Table 1.5. Although the antigenicity is very low in these concentrates, they still have antigenic properties in preruminant calves as low levels of soya-specific antibodies are formed (Tolman, 1991). In young piglets these concentrates appear to behave as being antigen-free (G.H. Tolman, unpublished results). Hydrolysing proteins by means of acid or proteases resulults in products apparently free of antigenic proteins. Lallès *et al.* (1991) showed that hydrolysed soya products were antigen- free according to the ELISA-test and did not stimulate the immune system in preruminant calves to produce circulating soya-specific IgG. In spite of the apparent absence of ANFs in the hydrolysate, the apparent protein digestibility of the product still was not as high as that of milk protein (0.82 *versus* 0.95).

α-AMYLASE INHIBITORS

The α-amylase inhibitor has been indicated as being responsible for the impaired digestion of starch from kidney beans (Jaffé and Vega Letta, 1968). *In vivo* experiments with either raw materials or purified inhibitors have clarified the marginal nutritional significance of α-amylase inhibitors. Red kidney bean inhibitor did not affect the growth of weanling rats, nor did it alter the availibility of energy from dietary starch (Savainao *et al.*, 1977). Therefore, this factor seems to be of minor importance.

TANNINS

Tannins are polyphenolic compounds able to form complexes with proteins, carbohydrates and other polymers in foods (Rao and Prabhavathi, 1982). Tannins form complexes more easily with proteins than with carbohydrates. This is related to the strong hydrogen-bond affinity to the carboxyl oxygen of the peptide group of proteins. Protein digestibility is reduced as a result of this complexation (Jansman *et al.*, 1989; Liebert and Gebhardt, 1983). The way dietary tannins act in the animal is not entirely clear. After consumption tannins may also form complexes with digestive enzymes resulting in a decreased enzyme activity. The latter may cause a decreased nutrient digestibility (Griffiths and Mosely, 1980; Marquardt 1989). Other antinutritional effects which may be related to tannins are damage to the intestinal mucosa, toxicity after absorption and interference with mineral availibility (Mitjavila *et al.*, 1977). Generally, tannins interfere with different aspects of the digestive processes resulting in reduced growth and increased feed conversion efficiency. In laying hens they also lower egg production (Marquardt, 1989).

Jansman *et al.* (1989) evaluated four cultivars of faba beans, differing in tannin content, in piglets and chickens at a level of 300 g/kg feed. The contents of tannins and other ANFs in these beans are summarized in Table 1.6.

In piglets, the four cultivars of faba bean were evaluated in an ileal and faecal digestibility study. Results were obtained with 6 piglets per treatment in a cross-over design. The weight of the piglets at the start of the experiment was 9 kg. The digestibility of the faba beans was calculated by difference. The ileal and faecal digestibilities of protein and N-free extract (NFE) are given in Table 1.7.

The results clearly show that in piglets the apparent ileal and faecal digestibility of crude protein decreases with increasing levels of tannins in the faba beans. Effects of TI and lectins are probably less important at these levels. Moreover, the batch with the lowest ileal protein digestibility contained the lowest levels of TIA and lectins. A lower apparent digestibility of faba bean protein in pigs due to the presence of tannins has also been reported by Liebert and Gebhardt (1983). NFE ileal apparent digestibility was

Table 1.6. LEVELS OF ANFs IN SAMPLES OF FOUR CULTIVARS OF FABA BEAN

Cultivar	*Flower colour*	*Tannins*[1] (g/kg)	*Trypsin inhibitors*[2] (mg/g)	*Lectins* (HA)[3]
Blandine	W	0.2	1.3	
Herz Freya	C	4.0	1.4	5
Mythos	C	9.8	1.6	5
Alfred	C	9.6	0.7	2

W = white; C = coloured.
[1] Measured according to the vanillin sulphuric acid method
[2] Expressed as mg trypsin inhibited per gram product
[3] Haemagglutination units with rabbit red blood cells; one HA is a 1:1000 dilution step.

about 0.200 lower than faecal apparent digestibility. This indicates that a substantial part of the NFE is digested in the large intestine, a process that is accompanied by relative large energy losses. The ileal apparent digestibility of NFE for the four cultivars of faba beans did not differ significantly. The faecal apparent digestibility of NFE appeared to decrease with increasing tannin content.

The same batches of faba bean cultivars were evaluated for three weeks in feeding trials with broilers. The experiment started when the birds were 5 days old. The beans were tested in experiments with isonitrogenous and isocaloric diets, balanced for digestible amino acid content (Jansman *et al.*, 1989). The results are summarized in Table 1.8.

There was no significant effect of diet on feed intake, weight gain or feed conversion efficiency (Table 1.8). The experiments with piglets and broilers show that piglets are more sensitive to tannins than chickens. This species-dependent sensitivity is in agreement with the results of Huisman *et al.* (1989, 1990b,c) who found that the piglet is also more sensitive to feeding Phaseolus beans and peas than chickens.

VICINE/CONVICINE

Vicine and convicine are present in faba beans. These agents are hydrolysed by the intestinal microflora (Frolich and Marquardt, 1983) to divicine and isouramil (Albano *et al.*, 1984). These metabolites cause haemolytic anaemia (favism) in man. Studies by Campbell *et al.* (1980), Olaboro *et al.* (1981) and Muduuli *et al.* (1982) demonstrated that vicine and convicine lead to a decrease in egg weight and egg size, weaker egg shells, an increased number of blood spots in the egg and a decreased fertility and hatchability of the eggs. Nielsen and Kruse (1974) found smaller numbers of piglets born per litter when

Table 1.7. ILEAL AND FAECAL APPARENT DIGESTIBILITY COEFFICIENTS OF CRUDE PROTEIN AND N-FREE EXTRACT (NFE) OF CULTIVARS OF FABA BEANS (INCLUSION LEVEL 300 g/kg FEED) IN PIGLETS

Cultivar	*Flower colour*	*Apparent digestibility coefficient*			
		Crude protein		*NFE*	
		ileal	*faecal*	*ileal*	*faecal*
Blandine	W	0.853[a]	0.893[a]	0.729[a]	0.943[a]
Herz Freya	C	0.753[b]	0.852[ab]	0.754[a]	0.930[ab]
Mythos	C	0.741[b]	0.824[bc]	0.727[a]	0.907[bc]
Alfred	C	0.687[b]	0.794[c]	0.691[a]	0.895[c]

W = white; C = coloured.
Values without a common superscript within a column differ significantly at $P<0.05$.

Table 1.8. FEED INTAKE, WEIGHT GAIN AND FEED CONVERSION EFFICIENCY OF BROILERS FED DIETS CONTAINING FABA BEANS DIFFERING IN TANNIN CONTENT

Diet	*Feed intake* (g)	*Weight gain* (g)	*Feed conversion efficiency* (kg feed/kg weight gain)
Control diet	1592	1039	1.53
30% Blandine in the diet	1547	1038	1.49
30% Herz Freya in the diet	1594	1047	1.52
30% Mythos in the diet	1600	1058	1.51
30% Alfred in the diet	1589	1046	1.52

Adapted from Jansman et al. (1989).

high levels of faba beans were fed to sows. The levels of vicine and convicine vary among the faba bean varieties. The activity of these compounds can be reduced chemically, but the costs of these treatments are generally too high for application in practice. Breeding varieties which are low in vicine and convicine is the most promising approach.

SAPONINS

Saponins are glycosides present in many plants. They are characterized by their bitter taste, their foaming properties in aqueous solution and their ability to haemolyse red blood cells (Birk and Peri, 1980). They are able to form complexes with sterols, including those associated with the plasma membranes of animal cells (Johnson *et al.*, 1986). Their antinutritional effect seems to be related to an increase of the permeability of the small intestinal mucosa cells, leading to inhibition of active nutrient transport and facilitating the uptake of components to which the gut normally would be impermeable (Johnson *et al.*, 1986). Various reports (reviewed by Birk and Peri, 1980) show that poultry are much more sensitive to saponins than other monogastric animals and ruminants. Levels of 20% lucerne meal in chick diets (equivalent to about 0.3% saponins) resulted in growth depression, which was attributed entirely to the saponin content. When the same levels of alfalfa were incorporated into pig diets no negative effects were found (Birk and Peri, 1980). For saponins in soya and other legumes these effects seem to be negligible. Generally, saponins can be considered as less important because levels are low in most common feed ingredients for monogastric animals.

FLATULENCE FACTORS

Flatulence factors are related to oligosaccharides which are fermented by intestinal bacteria in the large intestine. These oligosaccharides are not broken down in the small intestine because of a lack of appropriate enzymes, and subsequently enter the large intestine. There they are broken down by bacterial α-1,6-galactosidase. The monomers of these sugars are converted into volatile fatty acids, carbon dioxide, hydrogen and methane, resulting in flatulence, diarrhoea, nausea, cramps and discomfort (Rackis, 1975; Saini, 1989).

GLUCOSINOLATES

Glucosinolates are present in Cruciferae (Tookey *et al.*, 1980). Some 100 plant glucosinolates have been identified to date, seven of which are relevant for rapeseed. Glucosinolates *per se* seem to be non-toxic, but metabolites present after hydrolysis by myrosinase (thioglucosidase) are designated to have antinutritional effects. Myrosinase is present in the plant and in the seed, but also intestinal bacteria have appropriate enzyme systems for glucosinolate hydrolysis (Larsen, 1981). Because of the presence of myrosinase in the intestine, inactivation of myrosinase in the seed is not an appropriate way to eliminate the antinutritional effects of glucosinolates. Depending on pH, various compounds are formed through hydrolysis. Under neutral pH conditions oxazolidinethione (goitrins), thiocyanates and isothiocyanates are formed and under low pH conditions nitriles are produced (Larsen, 1981). The toxic effects of these compounds are described in various reports (Bell, 1984; Butler *et al.*, 1982; Henkel and Mosenthin, 1989; Pusztai, 1989; Rundgren, 1983; Tookey *et al.*, 1980). The main toxic effects of the hydrolytic products are depressed iodine uptake, thyroid hypertrophy and hyperplasia, changes in T3:T4 (tri-iodothyronine:thyroxine) ratio and liver damage. These effects lead to lower feed intake, poorer growth, raised feed conversion efficiency and, in poultry, decreased egg production. Although double-zero and even triple- zero cultivars of rapeseed are available, problems with rape seed still occur and hamper an unrestricted use of rapeseed meal, particularly for pigs and poultry. The precise role of the various glucosinolates is not entirely clear. Effects on thyroid size were found in pigs fed double-zero varieties in which the total glucosinolate content exceeded 5 μmol/g. This effect was particularly observed when the progoitrin content was above 3 μmol/g (Eggum *et al.*, 1985). In the same study thyroid hypertrophy was also observed with a double-zero variety (dietary inclusion level 200 g/kg) containing only 3 μmol/g. The use of rapeseed in poultry is limited in spite of the low glucosinolate levels in the double-zero cultivars. A serious problem is the liver haemorrhages in laying hens (reviewed by Fenwick and Curtis, 1980; Hill, 1979). This syndrome was still observed when rapeseed with a very low level of glucosinolates was fed, and even when the glucosinolates were removed (Wight *et al.*, 1987). It has been suggested that factors other than glucosinolates are the cause. This serious problem needs further research. Reduction of the negative effects of glucosinolates was found after addition of Fe, Cu and Mn to diets with 200 g/kg low- glucosinolate rapeseed (Menzel, 1983; Lüdke *et al.*, 1985; Schöne *et al.*, 1987).

Technological treatments to reduce the antinutritive effects of glucosinolates are generally insufficient and economically not feasible. More promising advantages can be expected from breeding new varieties low in erucic acid and glucosinolates. The European varieties (mainly winter varieties) generally contain higher levels of glucosinolates than the Canadian varieties (spring varieties). Therefore, insight into maximum inclusion levels of glucosinolates in diets of poultry need to be deepened. Based on this information plant breeders can be advised about the maximum tolerance levels for glucosinolates in rapeseed.

In the literature various recommendations for inclusion levels of rapeseed meal in diets for pigs and poultry are given. These recommended levels are variable depending on the varieties used (e.g. high- or low-glucosinolate cultivars, European or Canadian rapeseed; Butler *et al.*, 1982; Clandinin and Robblee, 1981; Henkel and Mosenthin, 1989). In Table 1.9 some recommended dietary inclusion levels of double zero rapeseed meal are given. However, no general recommendations about threshold levels for glucosinolates can be deduced from these reports.

ALKALOIDS

Alkaloids are bitter compounds present in many plants. Lupins contain the quinolizidine alkaloids (QA). Data on lupin alkaloids were reviewed by Cheeke and Kelly (1989). Alkaloids are oxidized in the liver, producing metabolites such as dehydrosparteine. Metabolic effects of alkaloids and its metabolites are primarily neural inhibition. Anti-palatability effects of lupin QA might be mediated in part by neurological effects (Cheeke and Kelly, 1989). Cheeke and Kelly (1989) have reported that dietary levels of lupins up to 25% can be used without effects on palatability and performance. Soya beans and linseed can be contaminated with *Datura stramonium L.* In Datura the alkaloids hyoscyamine and scopalamine are present, both of which affect the neural system. There is insufficient information about the toxicity of alkaloids in relation to animal nutrition. The literature

Table 1.9. RECOMMENDED MAXIMAL DIETARY INCLUSION LEVELS OF DOUBLE ZERO RAPESEED MEAL (g/kg AS FED)

Author	*1*	*2*	*3*	*4*	*5*	*Mean*
Piglets	50-80	-	-	<100	50	50-80
Fattening pigs	200	200	-	200	>200	200
Sows	<120	-	-	-	100	100
Broilers	<200	200	200	-	-	200
Laying hens	*	100	100	-	-	100

* Depending on sinapin level
1 Henkel and Mosenthin (1989)
2 Clandinin and Robblee (1981)
3 Butler, Pearson and Fenwick (1982)
4 Bell (1984)
5 Rundgren (1983)

suggests that sheep and rabbits are quite resistant to alkaloids from Datura, due to the presence of the enzyme atropine esterase. The pig is very sensitive, poultry are intermediate. The effects of scopolamine are more pronounced than those of hyoscyamine. There are very few data available on the maximum inclusion levels of alkaloids. With reference to the alkaloid concentration of Datura seed, the maximum content under EC law is limited to 1000 mg/kg for all Datura seeds.

PHYTATES

Phytic acid is a cyclohexane compound with 6 phosphate groups (inositol phosphate). The phosphate groups are capable of forming complexes with kations resulting in a reduced availability of Ca, Mg, Zn, Cu and Fe (Reddy *et al.*, 1982; Forbes and Erdman, 1983). Phytic acid has also been shown to interfere with the basic residues of proteins. Liener (1989) suggests that this may be a reason why phytate inhibits a number of digestive enzymes such as pepsin, pancreatin and α-amylase. More research is needed to investigate which aspect of enzyme inhibition can be related to the overall antinutritional effect. The utilization of phosphorus (P) is an important theme in discussions about environmental P pollution. The problem with P is mainly related to the poor digestibility of P in ingredients of vegetable origin. In these feed ingredients about 2/3 of the P is present as phytate P, which has a low digestibility in monogastric animals. Addition of the enzyme phytase to these ingredients can increase P digestibility considerably (Simons and Versteegh, 1990).

GOSSYPOL

Gossypol is the major pigment of cottonseed. The various aspects of gossypol have been reviewed in detail by Berardi and Goldblatt (1980). Depending on the varieties and environmental circumstances the content of gossypol may vary from 0.4 to 1.7%. The main physiological effects in poultry are loss of weight, decreased feed intake, decreased haemoglobin content, decreased egg size and decreased egg hatchability. Free gossypol can form complexes with proteins which may result in a decreased protein digestibility. Gossypol reacts strongly with chemicals (e.g. Fe salts). Fe salts can be used to reduce the negative effects of gossypol. Other possibilities of elimination are heat treatment and extraction. Today, new varieties of cotton seed with low levels of gossypol (<0.1 mg/kg) are cultivated.

SINAPINS

Sinapins are present in rapeseed at levels between 1.2 and 2.26%. Sinapins are converted in the large intestine to trimethylamine (TMA). TMA is responsible for taint (fishy odour) in eggs. Taint occurs at a sinapin level of >0.8 mg/kg (Fenwick, 1982). Taint is observed in brown-shelled eggs, but not in white-shelled eggs. Chemical treatment (alkali, calcium hydroxide and ammonia) in the presence of water reduces markedly the level of sinapins in rapeseed (Clandinin and Robblee; 1981). Another possibility of reducing the taint problem would be breeding of rapeseed varieties low in sinapins.

Some nutritional aspects of ANFs in the animal

Some aspects relating to the nutrition of animals and the effects of ANFs will be discussed briefly. For example, differences between animal species, threshold levels of ANFs, analysis and reduction of ANF activity.

ANIMAL SPECIES DIFFERENCES

Most reports dealing with the mode of action of ANFs in the animal and studies on the possibilities of reducing ANF activity by technological treatments have been carried out with small laboratory animals such as rats, chickens and mice. The reason for use of these is often related to the high costs and the large amounts of purified ANFs needed in studies with larger animal species. Criteria in these studies were often protein efficiency ratio (PER), net protein utilization (NPU) and N balance. Not only can the use of small laboratory animals as a model for larger animals be questioned but also the parameters used. It is uncertain whether PER, NPU and N balance values obtained in small animals, such as rats, can be extrapolated to larger farm animals like pigs. There are differences between small and larger animals in the reaction of the pancreas to trypsin inhibitors. Differences in response between rats and piglets to ANFs present in raw soya beans were suggested by Combs *et al.* (1967) and Yen *et al.* (1977). Visitpanich *et al.* (1985) found no difference in relative growth depression between rats and piglets fed pigeon pea (*Cajanus cajan*), but did find differences in response between rats and piglets when chick peas (*Cicer arietinum*) were fed. Huisman *et al.* (1990b,c) and Van der Poel *et al.* (1990a,b) found with rats and chickens on the one hand and piglets on the other hand differences in response to feeding *Phaseolus vulgaris*. Piglets showed distinctly more antinutritional effects to feeding raw *Phaseolus vulgaris* than rats and chickens. Greater antinutritional effects were found in piglets than in rats and chickens when peas (*Pisum sativum*) were fed (Huisman *et al.*, 1989). The results of Jansman *et al.* (1989) clearly showed that broilers were less sensitive to tannins than piglets. It was also demonstrated that at a more "practical" inclusion level, tannins of faba beans did not have negative effects in broilers. Therefore, with respect to the antinutritional effects of tannins reported in the literature, attention should be paid to inclusion levels and to differences among animal species. Summarizing, it is clear that the effects of ANFs measured in studies carried out with small laboratory animals cannot be extrapolated to larger farm animals; consequently, antinutritional effects should be studied in the target animal. Not only differences among animal species have to be taken into account, but also the age of animals. Piglets, for example, are more sensitive to trypsin inhibitors than fattening pigs.

THRESHOLD LEVELS

Until now hardly any attention has been paid to threshold levels, that is the levels of ANFs which can be tolerated without causing negative effects on performance of the animal. In the literature little indication about threshold levels can be found. This is related to the fact that most research has been focused on the mode of action in the animal using high levels of purified ANFs in the diets or high inclusion levels of ANF-containing seeds. The effect of ANFs is dependent not only on differences between animal species; the age of the animal may be important as well. Data about threshold levels for TIA have not been reported. Experiments at ILOB with broilers and fattening pigs have shown that adequately heated soya flour with a TIA level of 2-3 mg inhibited trypsin per g product can be included at dietary levels of 200 g/kg, without causing negative effects. This indicates that the maximum tolerance level in diets for broilers and fattening pigs is approximately 0.5 g TIA/kg diet. Different reports show that raw soya beans can be included in the diet of sows without causing negative effects (Crenshaw and Danielson, 1985). Summarizing, it seems that 0.5 g TIA/kg diet can be tolerated in diets for broilers, laying hens and fattening pigs. However, for young pigs this level may be too high because growth depression occurs at an inclusion level of 200 g/kg soya bean meal. It could be questioned whether the reduction of TIA in soya bean meal to a level of 2-3 mg TIA/g product is important or that the reduction in TIA serves as a marker for an adequate heat treatment necessary for the reduction of other factors such as lectins and for changes in protein structure to obtain a well digested protein. The results of Jansman *et al.* (1989) indicated that in chick diets tannin levels of up to 3 g/kg diet (when measured according to the vanillin sulphuric acid method) or 5 g/kg diet (when measured according the Folin Denis method) can be tolerated. However, the same study demonstrated that this tannin level decreased the apparent ileal protein digestibility in piglets. For other ANFs no threshold levels can be found. For lectins the situation is complicated. Various studies have shown that the toxicity of lectins varies. For example, lectins from *Phaseolus vulgaris* are very toxic, but those from peas are distinctly less toxic. This indicates that for lectins different threshold levels are to be used depending on type of lectins, animal species and age of the animals. As stated before, it is not possible to give threshold level for glucosinolates. There is a need to know more about threshold levels. This information is not only necessary for diet formulation, but it is also the basis for research focused on reduction of the negative effects of ANFs.

SOME ASPECTS OF ANALYSIS

Research into the mode of action in animals and into threshold levels is seriously hampered by the lack of adequate analytical methods. For example, there are various methods to determine the levels of tannins and other (poly)phenols, but none of these is specific enough and, moreover, no distinction can be made between toxic and non-toxic (poly)phenols. The standard trypsin inhibitor assay is based on the determination of TIs in soya. It is uncertain whether this assay can be used without adaptation to determine TI in other seeds. A point of consideration may be that bovine trypsin is mainly used for the determination of trypsin inhibitor activity in raw materials, also when these products were fed to pigs and poultry. Boisen (1989) and Liu and Markakis (1989) showed that the use of trypsin from different animal species in the assay gave different results. The use of bovine trypsin in the assay may therefore produce misleading results when these are correlated with biological effects in other target species. It is therefore recommended that trypsin from the target animal be used in assays. As mentioned before, different analytical methods have been employed and the units in which trypsin inhibitor activity is expressed often differ. Therefore, it is very difficult to compare results obtained in different laboratories. In this respect standardization of analytical methods for determining ANFs is important. *In-vitro* agglutination of red blood cells is commonly used as a method to detect and quantify lectins. This haemaglutination reaction is related to the binding of lectins to glycoconjugates of the red blood cells. Agglutination occurs due to binding of two erythrocytes to the same lectin molecule. Agglutination of blood cells is used as a measure for lectin activity. Huisman et al. (1990a), Jaffé *et al.* (1972), Jaffé (1980), Marquardt *et al.* (1975) and Newton and Hill (1983) have reported differences in agglutination activity as measured with red blood cells of different animal species. Huisman and Jansman (1991) have shown that soya bean lectins aglutinate red blood cells of rabbits, but not those of horses. A discussion point is that the aim of the lectin assay is to detect those lectins which have an affinity to the carbohydrates of the intestinal mucosa. One may discuss whether glycoproteins of red blood cells are identical to those in the gut wall. Evidence that red blood cells do not bind all lectins that bind to the gut wall has been reported by Kik (1991). She found that L-lectins from *Phaseolus vulgaris* agglutinate lymphocytes but display only weak activity towards red blood cells, while the E- lectins agglutinate red blood cells. However, both lectins were found to interfere with the gut wall. These findings show that the haemagglutination method is not sufficiently specific. An improved lectin assay is the ELISA method (Gabius, 1987). However, no separation between toxic and non-toxic lectins can be made using this method. A new approach in the ELISA technique has been developed by Hendriks *et al.* (1987) who coated microtitre plates with intestinal brush border membranes and demonstrated the binding of lectins to this matrix (ELISA-test). A very promising assay for the determination of different types of lectins and isolectins is the FLIA (functional lectin immuno assay), developed by Hamer *et al.* (1989). This method is based on the ability of lectins to bind to microtitre plates coated with either various carbohydrate matrices or brush border membranes. This method enables different (iso)lectins to be measured in a particular sample. This lectin assay is also very important for lectin inactivation studies. It may thus be possible for plant breeders to select new varieties based on the presence of toxic and non-toxic lectins.

A way to assay antigens is by ELISA. Either sera of parenterally immunized animals or of orally immunized animals can be used. Theoretically, the advantage of oral immunization is that serum antibodies are produced with the particular property of combining with those antigens that pass the mucosal barrier, whilst parenteral immunization induces the production of antibodies raised against all antigens present in the raw material. Therefore, antibodies produced after oral administration of antigens might be more selective and useful for determining the antigenicity of feedstuffs than antibodies produced after parenteral immunization. The orally induced antibodies need to be obtained from the animal that is supposed to receive the raw material that is being tested on antigenicity. This means that the antigenicity of raw materials intended for diets of piglets should be determined with antisera raised in piglets, whereas raw materials for calf diets undergo an ELISA with calf antisera. The advantage of the parenteral route of immunization is that antibodies can be obtained directed specifically against one of the proteins in the product (e.g. glycinin, β-conglycinin or lectin in soya products; legumin or vicilin in pea products). Because it is hardly possible to purify large amounts of specific proteins, the oral route of immunization is not feasible. The results of both types of assay give (semi-)quantitative information on antigenicity. A qualitative assay for detection of antigenicity is the immunoblotting method (Van Oort *et al.*, 1991). The method is based on separating the proteins by electrophoresis, followed by incubation with antisera. By using sera of orally immunized animals information can be obtained on the proteins in feedstuffs that are feed antigens.

SOME ASPECTS OF REDUCTION OF ANF-ACTIVITY

A major aspect of ANFs is the reduction of apparent protein digestibility. For the development of methods to reduce the activity of ANFs, it is important to know which part of the decrease in protein digestibility can be attributed to the increase of endogenous protein secretion, and which part can be related to a low digestibility of the protein as such. In a classic protein digestibility experiment the "apparent protein digestibility" is based on the excretion of undigested feed protein and endogenous protein. Using the 15N technique, in which the proteins in the animal are labelled, a distinction can be made between excreted undigested feed protein and endogenous protein. Using these figures the true protein digestibility of the feed protein can be calculated as well as which part of the low apparent protein digestibility is due to the excretion of endogenous protein. This information is important for studies aimed at reducing the negative effects of ANFs. For example, Huisman *et al.* (1991) found in peas an apparent ileal protein digestibility between 0.74 and 0.79. The true ileal protein digestibility of these batches of peas was between 0.93 and 0.95. These results indicate that the digestibility of the pea protein as such was high. The low apparent protein digestibility was caused by hypersecretion of endogenous protein. These results show that efforts to develop methods for increasing apparent pea protein digestibility has to be focused on factors causing hypersecretion of endogenous protein rather than on the enzymatic susceptibility of the protein. Huisman and Le Guen (1991) have demonstrated that pea ANFs are the factors that cause a reduction in apparent protein digestibility. This information is relevant to plant breeders because it shows that attention should be focused on ANFs and not on protein quality (in terms of digestibility). It also directs the biotechnologists in their attempts to develop enzymes with the aim of inactivating ANFs. In the example of peas the use of proteineases will not have a positive effect on the apparent protein digestibility, but the use of enzymes that break down ANFs are relevant. The use of ANF- ases may be interesting for those ANFs whose breakdown products are not toxic, such as trypsin inhibitors, lectins, antigenic protein and tannins. However, one has to be cautious with ANFs whose metabolites are toxic, such as glucosinolates, vicine and convicine.

References

Abbey, B.W., Norton, G. and Neale, R.J. (1979). *British Journal of Nutrition*, **41**, 39–45

Albano, E., Tomasi, A., Mannuzzu, L. and Arese, P. (1984). *Biochemical Pharmacology*, **33**, 1701–1704

Anderson, R.L., Rackis, J.J. and Tallent, W.H. (1979). In *Soy Protein and Human Nutrition*, pp. 209–234. Ed. H.L. Wilcke, D.T. Hopkins and D.H. Waggle. Academic Press, New York

Bell, J.M. (1984). *Journal of Animal Science*, **58**, 4, 996–1007

Berardi, L.C. and Goldblatt, L.A. (1980). In *Toxic Constituents of Plant Foodstuffs*, pp 183–237. Ed. I.E. Liener. Academic Press, New York

Bertrand, G., Séve, B., Gallant, B. and Tomé, R. (1988). *Sciences des Aliments*, **8**, 187–212

Birk, Y. (1961). *Biochimica et Biophysica Acta*, **54**, 378–381

Birk, Y. (1987). In *Hydrolytic Enzymes*, pp 257–305. Ed. A. Neuburger and K. Brocklehurst. Elsevier, Amsterdam

Birk, Y. (1989). In *Recent Advances of Research in Antinutritional Factors in Legume Seeds*, pp. 239–250. Ed. J. Huisman, A.F.B. van der Poel and I.E. Liener. Pudoc, Wageningen

Birk, Y. and Peri, I. (1980). In *Toxic Constituents of Plant Foodstuffs*, pp 161–182. Ed. I.E. Liener. Academic Press, New York

Bond, D.A. and Smith, D.B. (1989). In *Recent Advances of Research in Antinutritional Factors in Legume Seeds*, pp 285–296. Ed. J. Huisman, A.F.B. van der Poel and I.E. Liener. Pudoc, Wageningen

Boison, S. (1989). In *Recent Advances of Research in Antinutritional Factors in Legume Seeds*, pp 118–120. Ed. J. Huisman, A.F.B. van der Poel and I.E. Liener, Pudoc, Wageningen

Bowman, D. (1944). *Proceedings Society Experimental Biology Medicine*, **57**, 139–140

Borchers, R. (1961). *Journal of Nutrition*, **75**, 330–334

Borchers, R. (1962). *Journal of Nutrition*, **77**, 309–311

Brandt, H., Hong, K.C. and Glodek, P. (1985). *Zuchtungskunde*, **57**, 92–98

Broadway, R.M., Duffey, S.S., Pearce, G. and Ryan, C.A. (1986). *Entomologica Experimentalis et Applicata*, **41**, 33–38

Bush, R.S., Toullec, R., Caugant, I. and Guilloteau, P. (1991). In *New Trends in Veal Calf Production*, pp 253–256. Ed. J.H.M. Metz and C.M. Groenestein. Pudoc, Wageningen

Butler, E.J., Pearson, A.W. and Fenwick, C.R. (1982). *Journal of the Science of Food and Agriculture*, **33**, 866–875

Campbell, L.D., Olaboro, G., Marquardt, R.R. and Waddell, D. (1980). *Canadian Journal of Animal Science*, **60**, 395–405

Cheeke, P.R. and Kelly, J.D. (1989). In *Recent Advances of Research in Antinutritional Factors in Legume Seeds*, pp 189–201. Ed. J. Huisman, A.F.B. van der Poel and I.E. Liener. Pudoc, Wageningen

Cheeke, P.R. and Shull, L.R. (1985). *Natural Toxicants in Feeds and Poisonous Plants.* AVI, Westport, Connecticut

Chubb, L.G. (1982). In *Recent Advances in Animal Nutrition*, pp 21–37. Ed. W. Haresign. Butterworths, London

Clandinin, D.R. and Robblee, A.R. (1981). *Journal of the American Oil Chemists' Society*, **58**, 6, 682–686

Combs, G.E., Connes, R.G., Berry, T.H. and Wallace, H.D. (1967). *Journal of Animal Science*, **26**, 1067–1071

Crenshaw, M.A. and Danielson, D.M. (1985). *Journal of Animal Science*, **60**, 1, 163–170

Eggum, B.O., Just, A. and Sorensen, H. (1985). In *Advances in the Production and Utilization of Cruciferous crops*, pp 167–176. Ed. H. Sorensen. Martinus Nijhoff/Dr W. Junk, Dordrecht

Etzler, M.E. (1986). In *The Lectins, Properties, Functions and Applications in Biology and Medicine*, pp 371–435. Ed. I .E. Liener, N. Sharon and I.J. Goldstein. Academic Press, New York

Fenwick, G.R. (1982). *Proceedings Nutrition Society*, **41**, 277–288

Fenwick, G.R. and Curtis, R.F. (1980). *Animal Feed Science and Technology*, **5**, 255–298

Forbes, R.M. and Erdman, J.W. (1983). *Annual Review of Nutrition*, **3**, 213–231.

Frohlich, A.A. and Marquardt, R.R. (1983). *Journal of the Science of Food and Agriculture*, **34**, 153–163

Friedman, M. (1986). *Nutritional and Toxicological Significance of Enzyme Inhibitors in Foods.* Plenum Press, New York

Gabius, H.J., Engelhardt, R., Hellman, K.P., Hellman, T. and Ochsenfarht, A. (1987). *Analytical Biochemistry*, **165**, 349–355

Gallaher, D. and Schneeman, B.O. (1986). In *Nutritional and Toxicological Significance of Enzyme Inhibitors in Foods*, pp 167–185. Edited by M. Friedman. Plenum Press, New York

Gertler, A., Birk, Y. and Bondi, A. (1967). *Journal of Nutrition*, **91**, 358–370

Goldstein, I.J. and Poretz, R.D. (1986). In *The Lectins, Properties, Functions and Applications in Biology and Medicine*, pp 35–247. Ed. I.E. Liener, N. Sharon and I.J. Goldstein. Academic Press, New York

Green, G.M. and Nasset, E.S. (1983). *Journal of Nutrition*, **113**, 2245–2252

Green, G.M., Levan, V.H. and Liddle, R.A. (1986). In *Nutritional and Toxicological Significance of Enzyme Inhibitors in Foods*, pp 123–132. Ed. M. Friedman. Plenum Press, New York

Griffiths, D.W. (1979). *Journal of the Science of Food and Agriculture*, **30**, 458–462

Griffiths, D.W. and Mosely, G. (1980). *Journal of the Science of Food and Agriculture*, **31**, 255–259

Gumbmann, M.R., Spangler, W.L., Dugan, G.M., Rackis, J.J. and Liener, I.E. (1985). *Qualitas Plantanum. Plant Foods for Human Nutrition*, **35**, 275–315

Gupta, Y.P. (1987). *Plant of Human Nutrition*, **37**, 201–228

Hamer, R.J., Koninkx J.F.J.G., Van Oort, M.G., Mouwen J.M.V.M. and Huisman, J. (1989). In *Recent Advances of Research in Antinutritional Factors in Legume Seeds*, pp 30–33. Ed. J. Huisman, A.F.B. Van der Poel and I.E. Liener. Pudoc, Wageningen

Hasdai, A., Nitsan, Z., Volcani, R. and Birk, Y. (1989). *British Journal of Nutrition*, **62**, 529–537

Hendriks, H.G.C.J.M., Koninkx, J.F.J.G., Draaijer, M., van Dijk, J.E., Raaijmakers, J.A.M. and Mouwen, J.M.V.M. (1987). *Biochimica et Biophysica Acta*, **905**, 371–375

Henkel, H. and Mosenthin, R. (1989). *Ubersichten zür Tierernärung*, **17**, 139–190

Heppell, L.M.J., Sissons, J.W. and Banks, S.M.(1989). *Research in Veterinary Science*, **47**, 257–262

Hill, R. (1979) *British Veterinary Journal*, **135**, 3–16

Huisman, J. (1989). In *Nutrition and Digestive Physiology in Monogastric Animals*, pp 17–35. Ed. E.J. van Weerden and J. Huisman. Pudoc, Wageningen

Huisman, J. and Jansman, A.J.M. (1991). *Nutrition Abstracts and Reviews, Series B.* (In Press)

Huisman, J. and LeGuen, M.P. (1991). In *Digestive Physiology of the Pig*, pp 60–66. Ed. M.W.A. Verstegen, J. Huisman and L.A. Den Hartog. Pudoc, Wageningen

Huisman, J., Van der Poel, A.F.B. and Liener, I.E.(1989). *Proceedings of the First International Workshop on Antinutritional Factors(ANF) in Legume Seeds.* Pudoc, Wageningen

Huisman, J., Heinz, Th., Van der Poel, A.F.B., Van Leeuwen, P., Souffrant, W.B. and Verstegen, M.W.A. (1991). *British Journal of Nutrition.* (In press)

Huisman, J., Van der Poel, A.F.B., Kik, M.J.L. and Mouwen, J.M.V.M. (1989). *Journal of Animal Nutrition and Physiology*, **63**, 273–279

Huisman, J., Van der Poel, A.F.B., Mouwen, J.M.V.M. and Van Weerden, E.J. (1990c). *British Journal of Nutrition*, **64**, 3, 755–764

Huisman, J., Van der Poel, A.F.B., Van Leeuwen, P. and Verstegen, M.W.A. (1990b). *British Journal of Nutrition*, **64**, 3, 743–753

Huisman, J., Van der Poel, A.F.B., Verstegen, M.W.A., and Van Weerden, E.J. (1990a). *World Review of Animal Production*, **XXV** (2) 77–82

Jaffé, W.G. (1980). In *Toxic Constituents of Plant Foodstuffs*, pp 73–102. Ed. I.E. Liener. Academic Press, New York

Jaffé, W.G. and Vega Letta, C.V. (1968). *Journal of Nutrition*, **94**, 203–210

Jansman, A.J.M., Huisman, J. and Van der Poel, A.F.B. (1989). In *Recent Advances in Research of Antinutritional Factors in Legume Seeds*, pp 176–180. Ed. J. Huisman, A.F.B. van der Poel and I.E. Liener. Pudoc, Wageningen

Janzen, D.H., Juster, H.B. and Liener, I.E. (1976). *Science*, **192**, 795–796

Jindal, S., Soni, G. L. and Singh, R. (1982). *Journal of Plant Food*, **4**, 95–103

Johns, D.C. (1987). *New Zealand Journal of Agricultural Research*, **30**, 169–175

Johnson, I.T., Gee, J.M., Price, K., Curl, C. and Fenwick, G.R. (1986). *Journal of Nutrition*, **116**, 2270–2277

Kakade, M.L., Hoffa, D.E. and Liener, I.E. (1973). *Journal of Nutrition*, **103**, 1772–1778

Kardivel, R. and Clandinin, D.R. (1974). *Poultry Science*, **53**, 1810–1816

Khayamhashi, H. and Lyman, R.L. (1966). *Journal of Nutrition*, **89**, 455–464

Khorassani, G.R., Ozimek, L., Sauer, W.C. and Kenelly, J.J. (1989). *Journal of Animal Science*, **67**, 1634–1641

Kik, M.J.L. (1991). *Effects of Lectins in Legume Seeds on the Structure and Function of the Small Intestinal Mucosa.* PhD Thesis, State University, Utrecht, THE NETHERLANDS

Kik, M.J.L., Rojer, Mouwen, J.M.V.M., Koninkx, J.F.J.G., van Dijk, J.E. and van der Hage, M.H. (1989). *The Veterinary Quarterly*, **11**, 2, 108–115

Kilpatrick, D.C., Pusztai, A., Grant, G., Graham C. and Ewen, S.W.B. (1985). *FEBS Letters*, **185**, 299–305

Kunitz, M. (1945). *Science*, **101**, 668–669

Kunitz, M. (1946). *Journal Genetic Physiology*, **29**, 149–154

Lallès, J. P., Roger, L. and Toullec, R. (1991). In *Protein Metabolism and Nutrition, Volume 2*, pp 222–224. Ed. B.O. Eggum, S. Boisen, C. Borsting, A. Danfaer and T. Hvelplund. NIAS, Foulum

Larsen, P.O. (1981). In *The Biochemistry of Plants. A Comprehensive Treatise. Vol. 7, Secondary Plant Products*, pp 501–526. Edited by P.K. Strumpf and E.E. Conn. Academic Press, New York

Le Guen, M.P., Huisman, J. and Makkink, C.A. (1991a). In *Digestive Physiology of the Pig*, pp 207–210. Ed. M.W.A. Verstegen, J. Huisman and L.A. Den Hartog. Pudoc, Wageningen

Le Guen, M.P., Tolman, G.H. and Huisman, J. (1991b). In *Digestive Physiology of the Pig*, pp 99-103. Ed. M.W.A. Verstegen, J. Huisman and L.A. den Hartog. Pudoc, Wageningen

Liebert, F. and Gebhardt, G. (1983). *Archiv für Tierernärung*, **33**, 47–56

Liener, I.E. (1980). *Toxic constituents of plant foodstuffs*. Academic Press, New York

Liener, I.E. (1981). *Journal of the American Oil Chemists' Society*, **58**, 3, 406–415

Liener, I.E. (1986). In *The Lectins*, pp 527–552. Ed. I.E. Liener, N. Sharon and I.J. Goldstein. Academic Press, New York

Liener, I.E., (1989). In *Recent Advances of Research in Antinutritional Factors in Legume Seeds*, pp 6–13. Ed. J. Huisman, A.F.B. Van der Poel and I.E. Liener. Pudoc, Wageningen

Liener, I.E. and Kakade, M.L. (1980). In *Toxic Constituents of Plant Foodstuffs*, pp 7–71. Ed. I.E. Liener. Academic Press, New York

Liener, I.E., Deuel, H.J. and Fevold, H.L. (1949). *Journal of Nutrition*, **39**, 325–339

Liener, I.E., Nitsan, Z., Srisangnam, C., Rackis, J.J. and Gumbmann, M.R. (1985). *Qualitas Plantarum. Foods for Human Nutrition*, **35**, 243–257

Lüdke, H., Schöne, F. and Hennig, A. (1985). *Archiv Tierernährung*, **35**, 835–845

Liu, K. and Markakis, P. (1989). *Cereal Chemistry*, **66**, 415–422

Markakis, (1989).

Marquardt, R.R. (1989). In *Recent Advances of Research in Antinutritional Factors in Legume Seeds*, pp 141–155. Ed. J. Huisman, A.F.B. van der Poel and I.E. Liener. Pudoc, Wageningen

Marquardt, R.R., McKirdy, J.A., Ward, T. and Campbell, L.D. (1975). *Canadian Journal of Animal Science*, **55**, 421–429

Marquardt, R.R. and Ward, A.T. (1979). *Canadian Journal of Animal Science*, **59**, 781–789

Menzel, E. (1983). PhD Thesis, University of Kiel, Germany.

Miller, B.G., Newby, T.J., Stokes, C.R., Hampson, D., Brown, P.J. and Bourne, F.J. (1984). *American Journal of Veterinary Research*, **45**, 1730–1733

Miller, B.G., Bailey, M., Telemo, E. and Stokes, C.R. In *Proceedings of the Second Spring Conference Edinburgh*, pp 86-94. Ed. J.P.F.D'Mello and C.M. Duffus. Scottish Agricultural College, Edinburgh

Mitjavila, S., Lacombe, C., Carrera, G. and Derache, R. (1977). *Journal of Nutrition*, **107**, 2113–2121

Mowat, A.M. (1987). *Immunology Today*, **8** (3), 93–98

Muduuli, D.S., Marquardt, R.R. and Guenter, W. (1982). *British Journal of Nutrition*, **47**, 53–60

Naim, M., Gertler, A.and Birk, Y. (1982). *British Journal of Nutrition*, **47**, 281–288

Nesheim, M.C. and Garlich, J.D. (1966). *Journal of Nutrition*, **88**, 187–192

Newton, S.D. and Hill, G.D. (1983). *Nutrition Abstracts and Reviews, Series B*, **53**, 99–115

Nielsen, H.E. and Kruse, P.E. (1974). *Livestock Production Science*, **1**, (2), 179–185

Nitsan, Z. and Gertler, A. (1972). *Nutrition Metabolism*, **14**, 371–376

Nitsan, Z. (1991). In *Protein Metabolism and Nutrition, Volume 1*, pp 103–115. Ed. B.O. Eggum, S. Boisen, C. Borsting, A. Danfaer and T. Hvelplund, NIAS, Foulum

Nitsan, Z. and Alumot, E. (1965). *Poultry Science*, **44**, 1210–1214

Olaboro, G., Marquardt, R.R., Campbell, L.D. and Frohlich, A.A. (1981). *Journal of the Science of Food and Agriculture*, **32**, 1163–1171

Pistole, T.G. (1981). *Annual Review Microbiology*, **35**, 85–112

Pusztai, A. (1989). In *Recent Advances of Research in Antinutritional Factors in Legume Seeds*, pp 17–29. Ed. J. Huisman, A.F.B. Van der Poel and I.E. Liener. Pudoc, Wageningen

Rackis, J.J. (1975). In *Physiological Effects of Food Carbohydrates*, pp 207–222. Ed. J. Allen and J. Heilge. American Chemists Society, Washington DC

Rackis, J.J. and Gumbmann, M.R. (1981). In *Antinutrients and Natural Toxicants in Foods*, pp 203–237. Ed. R.L Orvy. Food and Nutrition Press, Westport, Connecticut

Rackis, J.J., Gumbmann, M.R. and Liener, I.E. (1985). *Qualitas Plantarum. Plant Foods Human Nutrition.* **35**, 213–242.

Rackis, J.J., Wolf, W.J. and Baker, E.C. (1986). In *Nutritional and Toxicological Significance of Enzyme Inhibitors in Foods*, pp 299–347. Ed. M. Friedman. Plenum Press, New York

Rao, B.S.N. and Prabhavathi, J. (1982). *Journal of the Science of Food and Agriculture*, **33**, 89

Read, J.W. and Haas, L.W. (1938). *Cereal Chemistry*, **15**, 59–68

Reddy, N.R., Sathe, S.K. and Salunkhe, D.K. (1982). *Advances in Food Research*, **28**, 1–92

Rogler, J.C. and Carrick, C.W. (1964). *Poultry Science*, **43**, 605–612

Roitt, I.M., Brostoff, J. and Male, D.K. (1985). *Immunology.* Churchill Livingstone, Edinburgh/Gower Medical Publishing, London, UK

Rundgren, M. (1983). *Animal Feed Science and Technology*, **9**, 239–262

Ryan, C.A. (1983). In *Variable Plant and Herbivores in Natural and Managed Systems*, pp 43–60. Ed. R.F. Denno and M.S. McClure. Academic Press, New York

Saini, H.S. (1989). In *Recent Advances of Research in Antinutritional Factors in Legume Seeds*, pp 329–341. Ed. J. Huisman, A.F.B. van der Poel and I.E. Liener. Pudoc, Wageningen

Savage, G.P. and Deo, S. (1989). *Nutrition Abstracts and Reviews, Series A*, **59**, (2), 66–68

Savaiano, D.A., Powers, J.R., Costello, M.J., Whitaker, J.R. and Clifford, A.J. (1977). *Nutrition Reports International*, **15**, 443–449

Saxena, H.C., Jensen, L.S., Spencer, J.V. and McGinnis, J. (1963). *Poultry Science*, **42**, 291–293

Schöne, F., Steinbach, G., Kirhner, E., Hennig, A. and Lüdke, H. (1987). *Acta Veterinaria*, **56**, 201–296

Simons, P.C.M., Versteegh, H.A.J., Jongbloed, A.W. Kemme, P.A., Slump, P., Bos, K.D. and Wolters, M.G.E. (1990). *British Journal of Nutrition*, **64**, 525–540

Sissons, J.W. (1989). In *Recent Advances in Animal Nutrition*, pp 261–282. Ed. W. Haresign and D.J.A. Cole. Butterworths, London

Sissons, J.W. and Tolman, H. (1991). In *Proceedings of the Second Spring Conference Edinburgh*, pp 62–85. Ed. J.P.F.D' Mello and C.M. Duffus. Scottish Agricultural College, Edinburgh

Sissons, J.W., Nyrup, A., Kilshaw, P.J. and Smith, R.H. (1982). *Journal of the Science of Food and Agriculture*, **33**, 706–710

Struthers, B.J., MacDonald, J.R., Dahlgrew, R.R. and Hopkins, D.T. (1983). *Journal of Nutrition*, **113**, 86–97

Summers, J.D., McConachie, J.D., Slinger, S.J. and Pepper, W.F. (1966). *Poultry Science*, **45**, 165–169

Tookey, H.L., VanEtten, C.H. and Daxenbichler, M.E. (1980). In *Toxic Constituents of Plant Foodstuffs*, pp 103–142. Ed. I.E. Liener. Academic Press, New York

Tolman, G.H., (1991). In *New Trends in Veal Calf Production*, pp. 241–246. Ed. J.H.M. Metz and C.M. Groenestein. Pudoc, Wageningen

Van der Poel, A.F.B. (1989). In *Recent Advances of Antinutritional Factors in Legume Seeds*, pp 213–229. Ed. J. Huisman, A.F.B. Van der Poel and I.E. Liener. Pudoc, Wageningen

Van der Poel, A.F.B. (1990a). Ph.D. Thesis. Agricultural University, Wageningen, The Netherlands

Van der Poel, A.F.B. (1990b). *Advances in Feed Technology*, **4**, 22–34

Van der Poel, A.F.B., Liener, I.E., Mollee, P.W. and Huisman, J. (1990b). *Livestock Production Science*, **25**, 137–150.

Van der Poel, A.F.B., Mollee, P.W., Huisman, J. and Liener, I.E. (1990a). *Livestock Production Science*, **25**, 121–135

Van Dijk, J.E. Fleddèrus, A., Mouwen, J.M.V.M. and Holzhauer, C. (1988). *Veterinary Research Communications*, **12** (1), 47–59

Van Oort, M.G., Hamer, R.J. and Tolman, G.H. (1991). In *New Trends in Veal Calf Production*, pp 247–252. Ed. J.H.M. Metz and C.M. Groenestein. Pudoc, Wageningen

Visitpanich, T., Batterham, E.S. and Norton, B.W. (1985). *Australian Journal of Agricultural Research*, **36**, 327–335

Webb, A.J. and Curran, M.K. (1986). *Livestock Production Science*, **14**, 415–423

Webb, J. (1989). *Pigs*, **5**, 20–25

Wight, P.A.L., Scougall, R.K., Shannon, D.W.F., Wells, J.W. and Mawson, R. (1987). *Research in Veterinary Science*, **43**, 313–319

Yannai, S. (1980). In *Toxic Constituents of Plant Foodstuffs*, pp 371–418. Ed. I.E. Liener. Academic Press, New York, US

Yen, J.T., Jensen, A.H. and Simon, J. (1977). *Journal of Nutrition*, **107**, 156–165

2

HOCK BURN IN BROILERS

S. A. TUCKER and A. W. WALKER

Agriculture Development and Advisory Service, Gleadthorpe, Meden Vale, Mansfield, Notts NG20 9PF

Industry background

Hock burn in broilers remains an issue of importance to the poultry industry in both economic and welfare terms. Despite the fact that our understanding of this complex phenomenon has improved considerably in recent years, there is no doubt that hock burn still affects many broilers to the detriment of both profitability and flock welfare.

The unsightly brownish-black lesions which appear not only on the hock but also on the breast and feet of broilers have been described collectively as contact dermatitis (McIlroy *et al.*, 1987). Histological examination of these lesions has revealed inflammation and necrosis of the epidermis, and in severe areas the damage can penetrate as far as the upper dermis (Lynn, Tucker and Bray, 1991). It is difficult to estimate what proportion of the U.K. flock may be affected by hock burn. A survey carried out in 1986 (McIlroy *et al.*, 1987) suggested the figure may be 20%. The financial implications of such a statistic will vary according to prevailing conditions in the marketplace and the purpose for which the birds are being grown. In times of over-supply the criteria for downgrading a damaged carcass may well be more stringent than at other times, and will apply particularly to birds grown for the oven-ready, rather than portioned, market.

Technical background

Applied research at a number of institutes, including Gleadthorpe, has enabled us to identify contributary factors, to describe their effect and to propose appropriate solutions to the hock burn problem. In the light of this it seems reasonable to suggest that the problem is due either to inefficient application of current knowledge and techniques, or to as yet unidentified causal agents.

The former may be the more likely explanation. The acute inflammation and necrosis typically seen in "burnt hocks," described above, is probably caused by prolonged contact with corrosive substances in the litter (Bray, 1984). The process of litter deterioration which precedes the development of these lesions is influenced by many factors, all of which have received rigorous attention from researchers (Lynn *et al.*, 1991). These factors include house environment, nutrition, equipment, management and disease. The objective of research to date has been to quantify the impact of these factors and hence be able to prescribe standards to which producers should adhere in order to reduce the risk of hock burn (Lynn *et al.*, 1991). In particular ADAS has strongly emphasised the importance of litter moisture content and litter surface friability. These two factors are central to the hock burn issue and hold the key to its solution. Work at Gleadthorpe showed that when these two factors combine to give high litter moisture and poor friability, then the risk of hock burn increased (Bray and Lynn, 1986). Lynn and Spechter (1987) showed that when litter moisture content exceeds 46% the litter surface will become wet and unfriable.

Effects of drinker design

Excessive litter moisture can be caused by over-consumption by the birds, or by spillage from the drinkers, or both. In the case of over-consumption nutritional factors may be implicated, but in either case drinker design and management can play a crucial part.

In an experiment at Gleadthorpe it was demonstrated that small cup designs reduced water consumption without affecting live weight gain. This meant drier, more friable litter and consequently a lower risk of hock burn. The experiment used twelve drinker systems:

1. Bell drinker (Plasson)
2. Bell drinker unballasted (Plasson)
3. Bell drinker (BEC watermaster 4)
4. Bell drinker unballasted (BEC 75)
5. Bell drinker (EB equipment)
6. Bell drinker (Rainbow Cavalier)
7. Nipple without drip cup (monoflo)
8. Nipple with drip cup (monoflo)
9. Auto-cup drinker (monoflo)
10. Small cup drinker (Swish)
11. Small cup drinker (Hart)
12. Long trough (Eltex)

Examination of the performance data from this experiment (Table 2.1) shows that drinker design had a significant effect on body weight at 49 days. The only treatment which depressed body weight was the nipple without drip cup. Differences in feed intake mirrored differences in body weight and although there were no significant differences in food conversion efficiency, there was a significant reduction in water usage on treatments 7, 8 and 9.

The ability of birds on the small cup and nipple drinker systems to achieve similar live weights to those on the bell drinkers whilst using less water indicates an improvement

Table 2.1: EFFECT OF DIFFERENT TYPES OF WATER DRINKER ON GROWTH RATE, FEED USAGE, WATER USAGE AND LITTER FRIABILITY IN BROILERS (FOR DETAILS OF DRINKER TYPE SEE TEXT)

	Treatment													
	1	*2*	*3*	*4*	*5*	*6*	*7*	*8*	*9*	*10*	*11*	*12*	*Mean*	*SED*
Mean body weight (kg) at 49 days	2.51	2.54	2.53	2.51	2.49	2.48	2.36	2.46	2.46	2.50	2.49	2.49	2.48	0.0389
Feed usage (kg/bird) Day old to 49 days)	4.91	4.93	4.93	4.89	4.84	4.92	4.54	4.79	4.81	4.84	4.88	4.88	4.85	0.0695
Water usage(l/bird) Day old to 49 days	8.62	8.68	8.54	8.54	8.64	8.36	7.37	7.68	8.13	8.64	8.71	8.76	8.39	0.1329
Litter friability score[1] at 45 days of age	4.7	5.0	4.3	4.7	4.7	3.8	2.5	3.2	3.2	4.5	4.0	4.8	4.1	0.5424

[1] Key to litter friability score

Score	Guidelines
1	Free flowing/crumbly; No capping in any area
2	Very slight capping just visible but mostly friable
3	Access to friable litter partially reduced (approx 50%)
4	Most areas capped but litter still friable in small areas
5	Extensive capping/crusting or compaction with access to friable litter negligible

in water conversion efficiency, which in this experiment was positively correlated with better litter friability (Table 2.1).

Having found a relationship between water consumption and friability, it remained to examine bird body condition in detail to assess any correlation between the litter condition scores and subsequent carcass damage. Five body condition criteria were selected for this analysis, namely: burnt hocks, breast blisters, burnt skin, feather follicle damage and dirtiness of feathering (Table 2.2). A sample of twenty birds from each of the 36 pens in the experiment was examined and for each pen a cumulative score was given for the five body condition categories.

Table 2.2. BODY CONDITION SCORING GUIDELINES

Category	*Score*	*Guidelines*
1. Burnt hock	1	No discolouration
	2	Slight discolouration
	3	Discolouration with small scab(s)
	4	Well established scab(s)
	5	Enlarged hock with large scab(s)
2. Breast blister	1	No thickening of skin
	2	Skin thickened but not actually blistered
	3	Blistered or thickened skin which would lead to downgrading after plucking
3. Burnt skin	1	No discolouration
	2	Some reddening in discrete areas
	3	Very red in discrete or merged areas
4. Feather follicle damage	1	No damage, normal follicles
	2	Some feathers missing or broken, slightly enlarged reddened follicles
	3	Feathering badly disrupted with broken, missing or rotting feathers, very swollen, reddened or infected follicles
5. Dirtiness of feathering	1	Clean feathers
	2	Some dirt
	3	Dirty
	4	Very dirty

The results of the body condition analysis showed that where a lower litter friability score was obtained, then carcass damage was reduced (Table 2.3). The treatments which produced the lowest friability scores also produced the lowest carcass damage scores. The correlation coefficient between the two measurements was 0.6794 ($P<0.05$).

Table 2.3. BODY CONDITION SCORE AT 48 DAYS

Treatment											
1	*2*	*3*	*4*	*5*	*6*	*7*	*8*	*9*	*10*	*11*	*12*
96.3	98.3	93.7	92.0	89.7	78.3	77.0	85.0	89.0	96.0	103.0	92.7

Mean = 90.9
SED = 4.654

The drinker study at Gleadthorpe concluded with two main practical points: first, some systems have features which reduce water usage without reducing growth rate (the small cup and nipple- and drip-cup designs); and second, water usage and litter moisture are linked to friability and carcass damage. Therefore systems which reduce water usage reduce downgrading.

Effects of nutrition

While Gleadthorpe's work on drinker design pointed the way for the development of systems which were less prone to spillage and which improved the bird's water conversion efficiency, it could not of course address the issue of over-consumption due to nutritional factors. This subject demanded the initiation of a separate programme of work which set out to investigate the nutritional causes of wet litter and the associated problems of greasy litter and high litter nitrogen content, both of which were suspected of contributing to the incidence of hock burn. Our work not only confirmed this suspicion but showed also that the effect of some feed ingredients on litter condition can be additive (Bray *et al.*, 1986).

Fat quality was the subject of recent investigation. Two fat qualities were compared on the basis of their unsaponifiable matter, oxidised fatty acid content, saturated and unsaturated fatty acid content. The two fats had the properties presented in Table 2.4.

In this experiment, by 39 days of age the effects of the fat quality could be clearly seen in the litter condition. The birds had been unable to utilise the poor fat successfully for weight gain and the litter had an ether extract content of 7.1% which was three times higher than the good quality fat treatment. Subjective scoring for the extent of wetness and greasiness showed that the good quality fat in the control diet produced a drier, more friable litter surface than the poor quality fat. Although there was no effect of fat quality on core litter moisture, the poorer surface friability was reflected in higher hock burn scores (Table 2.5).

Table 2.4. PROPERTIES OF GOOD AND POOR QUALITY FATS (%)

Property	*Good quality*	*Poor quality*
Unsaponifiable matter	1.9	14.1
Oxidised fatty acids	1.2	9.7
Fatty acid composition (%)		
Total saturated	34.1	45.8
Total unsaturated	65.9	54.2

Table 2.5. EFFECT OF FAT QUALITY ON LITTER CONDITION AND HOCK BURN AT 47 DAYS

Fat quality	*Litter Moisture*	*Friability*	*Wetness/ Greasiness*[1]	*Hock burn*	
				Males	*Females*
Good	45.49	4.00	3.74	3.3	2.7
Poor	44.70	4.53	4.03	3.6	3.1

[1] Key to Wetness/Greasiness Score

Score	Guideline
1	Dry, as at day old
2	Slightly damp/tacky
3	Damp/tacky. Sticks to bird's feet but some dry areas still accessible
4	Most areas wet/sticky/greasy (Bird's feathers likely to be soiled)
5	Soggy, squelchy or very wet/greasy. Leaves durable imprint when compressed or very slippery (could be on top of a cap).

Any dietary ingredient that increases the water consumption of birds will lead to an increase in litter moisture. Salt content and the quantity and quality of protein in broiler diets have recently been examined for their effect on litter condition. Two sodium levels were used in an experiment which also incorporated three amino acid levels and three protein quality levels. In addition to litter moisture the experiment examined the effect of these treatments on the nitrogen content of the litter surface (Tables 2.6 and 2.7).

Table 2.6. THE EFFECT OF PROTEIN QUALITY AND QUANTITY AT 2 SALT LEVELS ON THE MOISTURE LEVEL OF THE LITTER SURFACE AT 48 DAYS

Sodium (%)	*% lysine: % methionine*		
	1.10 : 0.40	1.26 : 0.45	1.46 : 0.53
0.129	41.99	45.50	47.21
0.267	53.01	53.55	53.27
	Protein Quality		
	Good	*Mixed*	*Poor*
0.129	42.54	44.02	48.14
0.267	53.64	53.36	52.83

Table 2.7. THE EFFECT OF PROTEIN QUALITY AND QUANTITY AT 2 SALT LEVELS ON THE NITROGEN CONTENT (%) OF THE LITTER SURFACE AT 48 DAYS

	% lysine: % methionine		
Sodium (%)	*1.10 : 0.40*	*1.26 : 0.45*	*1.46 : 0.53*
0.129	5.75	6.68	7.25
0.267	5.98	6.77	7.48
	Protein Quality		
	Good	*Mixed*	*Poor*
0.129	5.72	6.43	7.53
0.267	6.08	6.60	7.55

(Bray, 1985)

In the above experiment, protein quality was defined by protein source. Therefore a range of protein sources which were felt to be of high quality (de-hulled soya, full fat soya and herring meal) were compared with three which were felt to be of low quality (poultry by-product meal, soya 4417 and meat and bone meal (48% C.P.)). The intermediate or "mixed" quality treatments were formed of a 50:50 ratio of the high and low quality rations. Protein quantity was defined as excessive, adequate or deficient essential amino acid content. Using lysine and methionine as the first limiting amino acids, excessive quantities were defined as 20% above adequate requirement, and deficient quantities were defined as 20 adequate requirements. The sodium content was pitched firstly at a level which was considered the minimum at which live weight gain would not be reduced, and then at a level which was considered to be totally unlimiting to live weight gain (see Appendix A for the diet specifications).

The results of this experiment showed that the three factors of protein quality, quantity and salt level have a cumulative effect on litter condition when measured both in terms of surface nitrogen content and core moisture. When carcass quality assessments were made, it was found that the proportion of potentially down-graded birds increased by a factor of almost 6 when a diet high in salt, poor in quality and excessive in protein quantity was fed compared with a diet which was low in salt, of good quality and of adequate protein content (Figure 2.1). Intermediate diets produced down-grading figures between the two extremes. Suggested maximum figures for litter moisture, ether extract and nitrogen content are given in Table 2.8.

Table 2.8. LITTER MOISTURE, ETHER EXTRACT AND NITROGEN CONTENT MAXIMA (%): ADAS RECOMMENDATIONS FOR REDUCED RISK OF HOCK BURN IN BROILERS

Item	*Suggested maximum*
Core moisture	40
Surface moisture	40
Core ether extract	2.5
Surface ether extract	4.5
Surface nitrogen	5.5

It seems likely that the effect of diet quality on litter condition can be accounted for through water intake and hence water excretion. Excess salt must be removed, as must excess nitrogen. The latter can occur by over supply of essential amino acids and by the presence of excessive non-essential amino acids, when large quantities of poor quality protein are consumed.

Effects of environmental factors

Several environmental factors can influence litter moisture at a given dietary regime. The feed compounder's best efforts to adhere to good standards of nutritional formulation, in terms of protein content, amino acid balance and salt inclusion as described above, can be destroyed by physical factors acting in the broiler house. Condensation occurs on a surface (and that can include litter) when the temperature of that surface falls below the dew-point temperature, which in turn is determined by the moisture content of the air. One of the functions of insulation is to keep inside surface temperatures above the dew-point, which is one reason why thermal conductance of better than 0.5 $W/m^2.{}^\circ C$ is

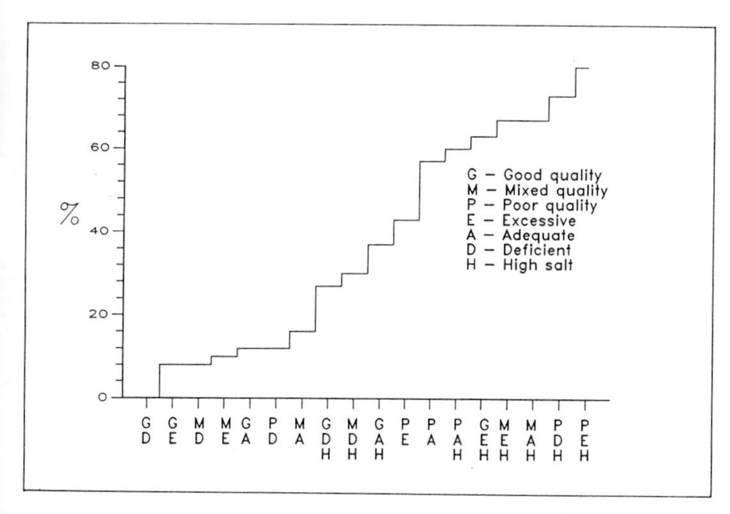

Figure 2.1. The effect of protein quality and quantity at 2 salt levels on % hock burn (Score 3 or above)

the traditional recommendation in the U.K. The effect of house temperature and relative humidity on dew-point temperature is shown in Figure 2.2. This clearly shows that the risk of condensation is higher when house temperatures are low and relative humidity is high.

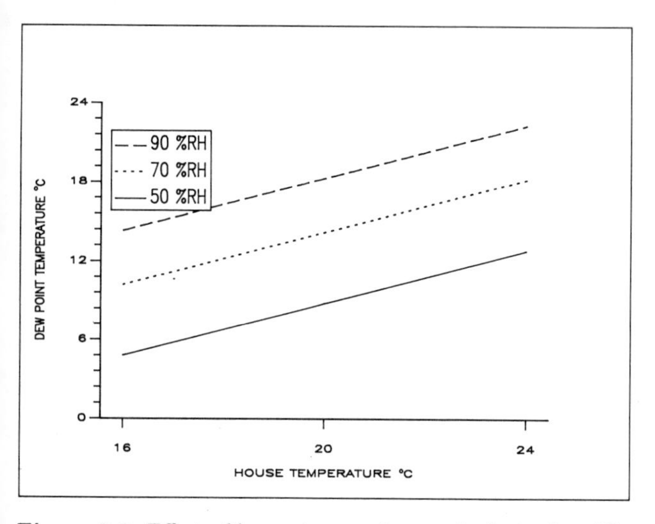

Figure 2.2. Effect of house temperature and relative humidity on dew-point temperature

It is in situations where house temperatures fall that litter management becomes difficult. In cold weather producers are tempted to reduce ventilation rates to conserve heat, but since this profoundly affects air moisture content it is vital that minimum ventilation rates are maintained. The minimum rate has traditionally been calculated as the amount of air necessary to prevent carbon dioxide levels exceeding about 0.3%, and this means approximately 1.6 $m^3/s \times 10^{-4}$ per kg liveweight$^{0.75}$ (Charles, 1981). To maintain this rate in cold weather while keeping house temperature at about 21°C almost certainly means using more fuel for heat. But the benefit from providing this extra heat, in terms of better litter and fewer down-grades, is quite likely to outweigh the cost of the fuel (Table 2.9).

Table 2.9. EFFECT OF BIRD RESPIRATION ON AIR MOISTURE BALANCE - THE BENEFIT OF EXTRA AIR SUPPLY

Age (days)	*Live weight* (kg)	*Minimum ventilation rate* 1.6 $m^3/s \times 10^{-4}$ per kg $M^{0.75}$		3.2 $m^3/s \times 10^{-4}$ per kg $M^{0.75}$	
		House relative humidity (%)	*Dew point* (°C)	*House relative humidity* (%)	*Dew point* (oC)
47	2.23	57	11.9	50	9.8
49	2.68	57	11.9	50	9.8

In the example shown in Table 2.9, the following assumptions have been made:- that the thermal conductance of walls and roof is better than U = 0.5 W/m^2 °C, that outside temperature is 10°C and Relative humidity 90%, and that the inside temperature is being held at 21°C by the brooder thermostat. Stocking density is assumed to be approximately 20 birds/m^2. It can be seen that an increase in ventilation rate has decreased the house Relative humidity and in doing so has shifted dew point temperature down a valuable 2.1°C, hence reducing the risk of condensation. However, many broiler houses have combinations of deteriorated insulation and imprecisely controlled ventilation rate which make the full exploitation of this phenomenon impossible. Payne (1967) found a correlation between litter quality and the air relative humidity recorded the previous week. Mean weekly relative humidities above 72% were associated with poor litter.

Houses which suffer from rising damp, condensation falling onto the litter from pipework or internal surfaces, or excessive water consumption from causes already described, are unlikely to benefit fully from this type of heating regime. These environmental deficiencies must first be put right, and in particular attention should be paid to the design and operation of the ventilation system. The controllability of the ventilation rate and the uniformity of air distribution must be optimised, otherwise the fuel costs of achieving the improvement in environment described above may be excessive.

A detailed discussion of ventilation systems design is outside the scope of this review. However, a field monitoring technique has been described by Sutcliffe, King and Charles (1987) which has allowed some conclusions to be drawn about the efficacy of ventilation systems.

Effect of litter materials

The behaviour of different litter materials under the moisture load imposed by a flock of broilers has implications for hock burn. Work at Gleadthorpe (Lynn and Spechter, 1986) investigated four materials:- woodshavings, paper, barley straw and wheat straw in terms of their effect on surface friability, performance and carcass quality. The experiment used three depths of each material: 2.5, 5.0 and 10.0 cm.

Table 2.10. LITTER SURFACE FRIABILITY SCORE AT 20, 41 AND 48 DAYS

Litter depth (cm)	*Litter material*				*Mean*
	Wood	*Paper*	*Barley straw*	*Wheat straw*	
20 days					
2.5	2.33	4.00	3.67	3.33	3.33
5.0	2.00	3.67	3.67	3.00	3.08
10.0	2.00	2.33	3.00	2.33	2.42
Mean	2.11	3.33	3.44	2.89	2.94
41 days					
2.5	3.33	3.67	3.67	3.67	3.58
5.0	3.00	3.67	3.33	3.00	3.25
10.0	2.00	4.00	4.00	3.00	3.25
Mean	2.78	3.78	3.67	3.22	3.36
48 days					
2.5	4.00	4.00	4.00	4.00	4.00
5.0	3.67	3.67	4.00	4.00	3.83
10.0	3.67	4.00	4.00	3.67	3.83
Mean	3.78	3.89	4.00	3.89	3.89

Litter friability at 48 days was poor on all treatments (Table 2.10), possibly exacerbated by humid weather conditions, but when friability was measured earlier in the flock, at 20 and 41 days, there was a trend for woodshavings to produce the most friable surface. The effect of litter depth on friability was more marked in the early stages. By day 20, paper litter at 2.5 cm depth was completely capped. By contrast, woodshavings at 10 cm depth remained friable until 41 days of age.

As found in earlier Gleadthorpe work described in this review, there was a positive correlation between increasing litter moisture content (Table 2.11) and poorer litter condition on each sampling occasion. The extent of capping, as indicated in the friability scoring, and litter moisture content, were positively correlated with the degree of hock burn (Tables 2.12 and 2.13) when measured at 49 days.

Male birds recorded the lowest hock burn scores at 49 days on wood-shavings at a depth of 10 cm. Female birds recorded low hock burn scores on both wood and wheat

Table 2.11. LITTER CORE MOISTURE CONTENT (%) AT 41 AND 48 DAYS

Litter depth (cm)	*Litter material*				*Mean*
	Wood	*Paper*	*Barley straw*	*Wheat straw*	
41 days					
2.5	50.03	53.97	51.83	52.67	52.13
5.0	47.07	52.60	52.10	49.23	50.25
10.0	39.80	48.60	49.83	45.60	45.96
Mean	45.63	51.72	51.26	49.17	49.44
48 days					
2.5	51.37	53.60	52.57	52.50	52.51
5.0	48.63	53.40	52.03	49.20	50.82
10.0	40.80	50.17	50.27	43.37	46.15
Mean	46.93	52.39	51.62	48.36	49.83

Effect of depth at 41 and 48 days $P<0.001$
Effect of material at 41 and 48 days $P<0.001$

Table 2.12. HOCK BURN SCORES AT 49 DAYS IN MALES AND FEMALES

Litter Depth (cm)	*Litter material*				*Mean*
	Wood	*Paper*	*Barley straw*	*Wheat straw*	
Males					
2.5	2.47	2.33	2.07	2.40	2.32
5.0	2.07	2.37	2.27	2.23	2.23
10.0	2.10	2.50	2.27	2.43	2.32
Mean	2.21	2.40	2.20	2.36	2.29
Females					
2.5	2.27	2.10	2.07	2.17	2.15
5.0	2.03	2.23	1.97	1.77	2.00
10.0	1.70	2.03	2.00	1.83	1.89
Mean	2.00	2.12	2.01	1.92	2.01

straw treatments, with wheat straw giving a marginally lower score overall. It was concluded that for the lowest litter moisture, most friable litter surface and the best quality birds, woodshavings should be used at a depth of 10 cm.

Table 2.13. CORRELATION COEFFICIENTS BETWEEN PARAMETERS AT 41 AND 48 DAYS

	41 days		48 days	
Litter surface friability vs litter moisture	0.7984	(*P*<0.01)	0.6095	(*P*<0.05)
Litter moisture vs hock burn	0.6202	(*P*<0.05)	0.6268	(*P*<0.05)
Litter surface friability vs hock burn	0.6751	(*P*<0.05)	0.3020	(NS)

Although alternatives to woodshavings are favoured by some broiler producers, this study suggests that the potential penalties for getting the litter management wrong, particularly at shallow depths, could be high. Nevertheless producers should take into account prevailing price and supply conditions in the market before ruling out the alternatives.

Effect of stocking density

Experimental work undertaken at Gleadthorpe during 1991 investigated the effect of stocking density on broiler welfare. Broilers were grown at a range of stocking densities: 12.2, 14.4, 15.8, 17.6, 19.8 and 22.7 birds/m^2. Hockburn score was significantly higher with increasing stocking density for males at 35, 43 and 50 days of age (Figure 2.3).

At both the lowest stocking density (12.2 bird/m^2) and the highest (22.7 birds/m^2), the hock burn score increased with age, but was always greater at the higher stocking density. This stocking density effect is shown across the range of treatments (Figure 2.4) not simply at the extremes. The number of birds with a high score (3 or above) increased with increasing stocking density.

The greater incidence of high hock burn scores on the birds grown at the higher stocking densities can be explained by the litter friability scores. Increasing stocking density within the range 12.2 to 22.7 birds/m^2 adversely affected litter friability (Table 2.14).

Litter friability was significantly poorer at the higher stocking densities from 21 days of age and deteriorated more rapidly at the higher densities. An explanation for the poorer litter condition can be found by examination of the total bird weight per m^2 of floor area (Figure 2.5).

At a stocking density of 22.7 birds/m^2, one square metre of litter must absorb the waste products of 57.22 kg of bird mass, whereas at the lower stocking density of 12.2 birds/m^2, the same area of litter absorbs the waste product of only 33.76 kg of bird mass,

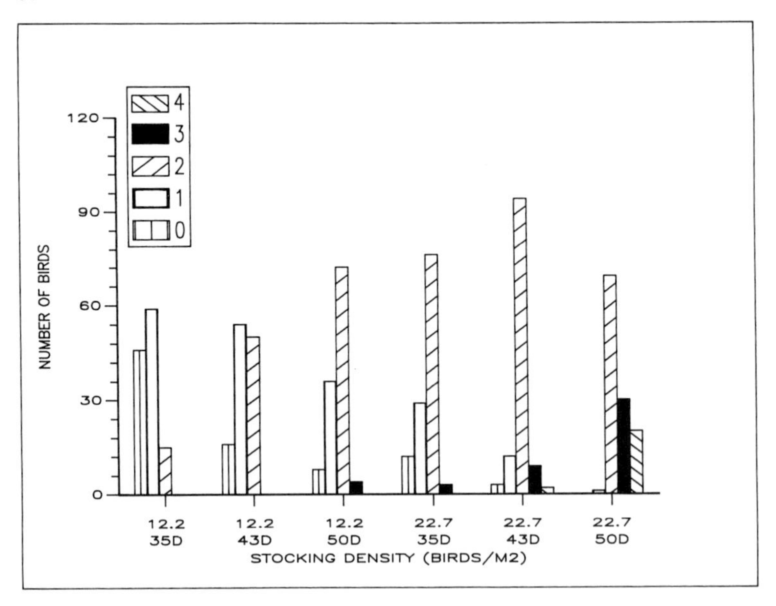

Figure 2.3. Effect of age and stocking density on the hock-burn score of male broilers at different ages

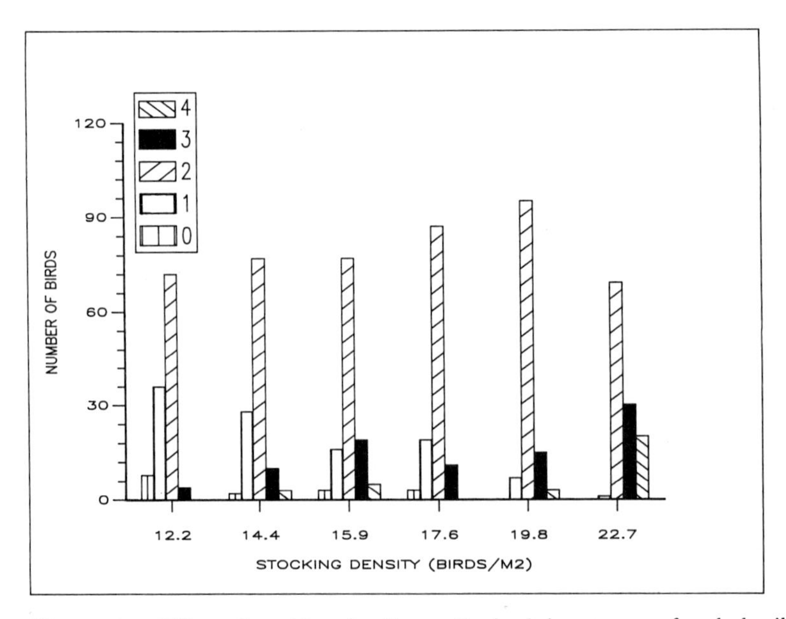

Figure 2.4. Effect of stocking density on the hock-burn score of male broilers at 50 days of age

Table 2.14. THE EFFECT OF STOCKING DENSITY ON LITTER FRIABILITY

Age (days)	*Stocking density* (birds/m^2) *12.3*	*14.5*	*16.0*	*17.8*	*20.0*	*22.9*	*Mean*	*SED*	*P*
21	1.75	2.25	3.25	2.75	3.75	4.00	2.96	0.296	<0.001
41	1.75	3.75	4.00	4.00	4.75	5.00	3.87	0.371	<0.001
48	3.00	4.00	4.50	4.75	5.00	5.00	4.37	0.314	<0.001

a reduction of 41%. Common sense suggests that if the bird numbers per unit area of litter is increased this will affect the critical factors of litter moisture, litter nitrogen and litter ether extract.

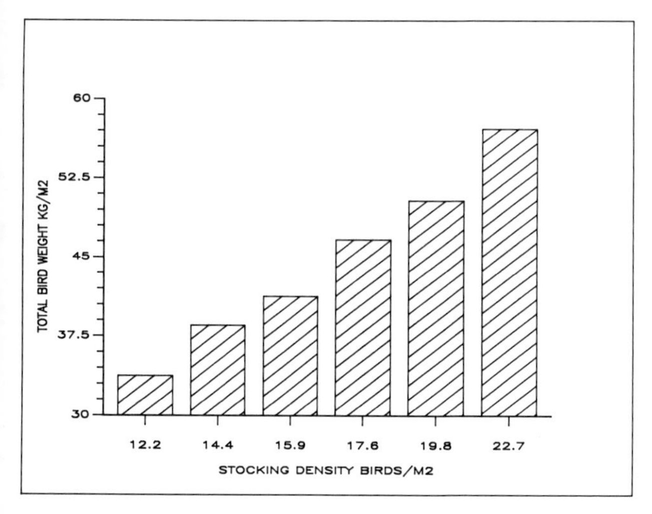

Figure 2.5. Effect of stocking density on the total live weight (Kg/m^2) of 49-day old broilers

Both core and surface litter moisture significantly increased with increasing stocking density at both 41 and 48 days of age (Table 2.15). Likewise the surface litter nitrogen content and litter ether extract content both significantly increased with stocking density (Table 2.16).

Conclusions

Despite the complexity of the hoch burn phenomenon, the principal causal mechanisms have been described. Responses to several important determinants of hock burn have been quantified. Of particular importance is the correlation between the incidence of hock damage and litter moisture and surface friability. Commercial operators can take effective measures to combat the problem, but these are likely to be multidisciplinary and there is, therefore, no simple panacea.

Table 2.15. SUMMARY OF LITTER MOISTURE CONTENT (%)

Age days	*Stocking density*(birds/m^2) 12.2	14.4	15.9	17.6	19.8	22.7	*Mean*	*SED*	*P*
Core									
41	32.9	37.8	40.4	42.9	42.8	47.8	40.8	2.585	<0.001
48	33.6	39.7	39.4	43.4	48.1	50.9	42.5	2.264	<0.001
Surface									
41	33.3	40.5	40.1	40.8	45.5	48.9	41.5	2.677	<0.001
48	35.8	42.5	43.2	43.1	49.7	54.6	44.8	3.093	<0.001

Table 2.16. SUMMARY OF LITTER SURFACE NITROGEN AND ETHER EXTRACT CONTENTS (%)

Age days	*Stocking density* (birds/m^2) 12.2	14.4	15.9	17.6	19.8	22.7	*Mean*	*SED*	*P*
Nitrogen content									
41	4.52	4.75	5.15	5.25	5.26	6.09	5.17	0.229	<0.001
48	5.10	5.58	5.76	5.73	6.30	6.37	5.81	0.149	<0.001
Ether extract content									
48	2.30	2.93	2.67	2.65	3.30	3.85	2.95	0.273	<0.001

In order to minimise the risk of hock burn, standards of environment control, nutrition and management must receive careful attention. In particular, temperature controllability must be optimised (Sutcliffe *et al.*, 1987) and ventilation rates maintained at those specified by Charles (1981). The risk of condensation will be reduced when house relative humidity is kept below 72% (Payne, 1967), and the moisture load on the litter reduced by the use of efficient drinker systems (Lynn and Spechter, 1987).

Nutritional factors have been shown to affect water consumption and care should be taken, when formulating diets, that salt and protein content do not exacerbate excessive water usage (Bray and Lynn, 1986). Where protein and fat are in excess, high litter nitrogen and ether extract content will also contribute to the risk of hock damage. High stocking densities have been shown to increase the incidence of hock burn because of the increased moisture and nitrogen load on the litter. As this relationship is linear, it may be necessary to reduce current commercial densities.

Acknowledgements

The authors acknowledge the work of T.S. Bray and N.J. Lynn, Gleadthorpe, and the financial support of the Ministry of Agriculture, Fisheries and Food.

References

Bray, T.S. (1984). *The effect of the diet on the litter condition and downgrading of broilers.* ADAS internal report, Gleadthorpe EHF, Meden Vale, Mansfield, Notts NG20 9PF

Bray, T.S. and Lynn, N.J. (1986). Effects of nutrition and drinker design on litter condition and broiler performance. *British Poultry Science*, **27** (1), 151

Charles, D.R. (1981). Practical ventilation and temperature control for poultry. In: *Environmental aspects of housing for animal production.* Ed. Clark J.A. Butterworths, London

Lynn, N.J., Tucker, S.A. and Bray, T.S. (1991). Litter condition and Contact Dermatitis in broiler chickens. In *Quality of Poultry Products Poultry meat.* Proceedings of Spelderholt Jubilee Symposia, Doorwerth, Netherlands, May 1991. Ed. T.G. Vijttenboogaart and C.H. Veerkamp

Lynn, N.J. and Spechter, H.H. (1986). *The effect of litter material and depth on broiler performance and aspects of carcass quality.* FAC Report No. 497, Gleadthorpe EHF, Meden Vale, Mansfield, Notts, NG20 9PF

Lynn, N.J. and Spechter, H.H. (1987). *The effect of drinker design on broiler performance, water usage, litter moisture and atmospheric ammonia.* FAC report No. 488, Gleadthorpe EHF, Meden Vale, Mansfield, Notts NG20 9PF

McIlroy, S.G., Goodall, E.A. and McMurray, C.H. (1987). A contact dermatitis of broilers - epidemiological findings. *Avian Pathology*, **16** (1), 93-105

Payne, C. G. (1967). Factors influencing environmental temperature and humidity in intensive broiler houses during the post brooding period. *British Poultry Science*, **48**, 1297-1303

Sutcliffe, N.A., King, A.W.M. and Charles, D.R. (1987). Monitoring poultry house environment. In *Computer Applications in Agricultural Environments.* Ed. Clark, J.A., Gregson, K. and Saffell, R. Butterworths, London

Appendix A

Experimental diets. Grower diets from 18 - 49 days (kg/tonne)

Protein quality:-		*Good*			*Poor*	
Amino acid levels:-	*Deficient*	*Adequate*	*Excessive*	*Deficient*	*Adequate*	*Excessive*
Ingredients						
Wheat	746	640	562	680	575	488
Extracted Soya (44%)	-	-	-	172	230	264
De-hulled Soya (48%)	133	160	188	-	-	-
Full fat Soya	-	80	130	-	-	-
Herring meal (70%)	30	30	30	-	-	-
Poultry by-product Meal	-	-	-	40	80	120
Meat and Bone meal (48%)	-	-	-	20	40	60
Hydrolised Edible fat	50	50	50	50	50	50
Dried skim milk	10	10	10	10	10	10
Limestone flour	10	10	10	8	5.5	-
Di-Calcium Phosphate	14	13	12	10	3.0	-
Salt	1.6	1.7	1.7	2.1	1.6	1.0
Sodium bicarbonate	1.1	1.0	1.0	-	-	-
Min./Vit. Supplement (containing 600g/tonne methionine)	5.0	5.0	5.0	5.0	5.0	
Methionine	0.260	0.800	1.500	0.400	0.800	1.400
Sacox	1.0	1.0	1.0	1.0	1.0	1.0
Eskalin 200	0.500	0.500	0.500	0.500	0.500	0.500
Calculated analysis (g/kg)						
ME(MJ/kg)	13.27	13.44	13.48	13.15	13.18	13.28
Crude Protein	167	199	223	180	227	259
Lysine	8.5	10.9	12.8	8.5	10.9	12.8
Methionine	3.8	4.8	5.8	3.8	4.8	5.8
Methionine + Cystine	6.6	8.0	9.4	7.2	9.3	1.1
Ether extract	6.7	8.0	8.8	7.7	8.9	10.2
Calcium	9.2	9.1	9.0	9.2	9.2	9.0
Total phosphorus	6.4	6.4	6.4	6.4	6.4	7.1
Avail. phosphorus	4.4	4.2	4.1	4.1	4.1	4.6
Sodium	1.3	1.3	1.3	1.3	1.3	1.3
Potassium	6.0	7.3	8.4	6.4	7.4	7.8
Chloride	2.0	2.0	2.0	2.2	1.9	1.9
For high sodium diets:- Add 2.3 kg salt						
Sodium	2.2	2.2	2.2	2.2	2.2	2.2
Chloride	3.4	3.4	3.4	3.6	3.4	3.3

3

PROTEIN QUALITY AND AMINO ACID UTILISATION IN POULTRY

K.N. BOORMAN

University of Nottingham School of Agriculture,
Sutton Bonington, Loughborough, LE12 5RD, UK

Introduction

The quality of a dietary protein source is defined as the extent to which the source can meet the essential amino acid needs of the animal. Quality can therefore be resolved into digestibility and the capacity of the absorbed amino acids to meet tissue needs, i.e. the concordance between the pattern of absorbed essential amino acids and the pattern of tissue requirements. There is rekindled interest in the relationship between the most deficient (limiting) amino acid in the diet and the quality of the protein mixture from which it is supplied. This has arisen from a realisation of the diversity of protein concentrates which may have to be used throughout the world, concern about nitrogen in effluents from poultry enterprises and argument about the validity of empirical methods used to measure amino acid requirements of poultry. The subject raises questions about maximising the responses to proteins of different quality and modifying protein quality by supplementation with free (synthetic) amino acids. Central to these topics is the utilisation of dietary amino acids. These subjects are the themes of this paper.

Quality and the theoretical response to protein

In protein-limiting circumstances, protein deposition (e.g. growth or egg production) will increase more or less linearly as dietary protein (protein intake) increases, until a limit (plateau) in deposition is reached (Figure 3.1). In terms of the growing chicken the limits of the response are: 1. at the lower end, the weight loss associated with zero protein intake, and 2. at the upper end, the maximum gain achievable. In theory these limits are functions of the bird and the environment and are therefore independent of protein quality. Protein quality expresses itself as the slope of the linear phase of the response curve. The smaller the slope (i.e. the more protein required to achieve a given gain) the poorer the quality of the protein (Figure 3.1). Allison (1964) pointed out that this slope represents Net Protein Ratio – a classical coefficient of protein quality. The equivalent relationship between nitrogen retention and nitrogen intake yields a slope representing Net Protein Utilisation.

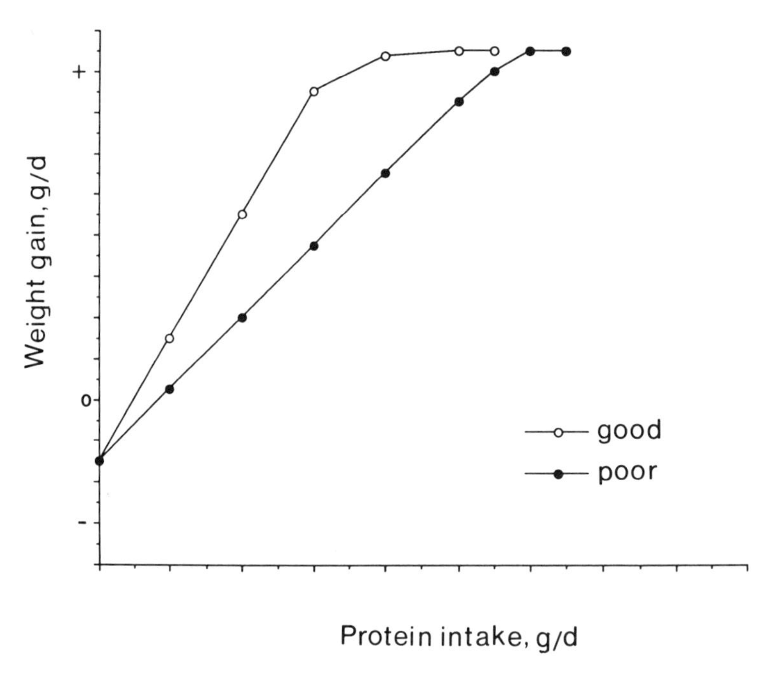

Figure 3.1. The theoretical responses in weight gain to intakes of proteins of good and poor quality

The theory of protein response therefore suggests that provided enough of a protein can be fed, maximum gain can be achieved, irrespective of the quality of the protein. Is there evidence to support this suggestion?

Response to poor-quality proteins

In respect of protein, Carpenter and de Muelenaere (1965) probably first posed the question: 'will quantity make up for quality...?' They used groundnut flour as the sole source of dietary protein for young chickens. They noted that although methionine is the most deficient (first limiting) amino acid in groundnut protein, severe processing can damage lysine sufficiently to cause this amino acid to become first limiting. Accordingly, they used a groundnut flour - lysine mixture to ensure primary limitation in methionine (a Chemical Score of about 50%). The mixture was fed at increasing concentrations in the diet with or without methionine supplementation. Selected results (Table 3.1) show that gain was maximised at concentrations of methionine of 3.5 g/kg and more and that this methionine could be provided solely from protein or from protein and synthetic methionine. This indicates that methionine requirement can be satisfied by increasing the concentration of a poor-quality protein. The authors thoroughly reviewed other available evidence, including that for rats, concluding cautiously that: 'under certain conditions, higher levels of poor protein will result in nearly as good growth as can be obtained with practical diets containing good quality protein'.

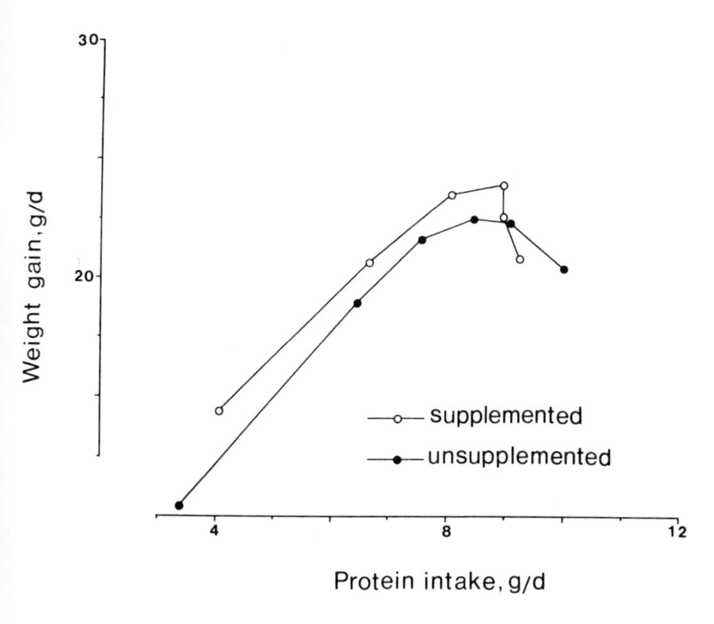

Figure 3.2. Weight-gain responses of young chickens to intake of protein from soyabean-maize mixtures unsupplemented or supplemented with methionine. Calculated from Wethli *et al.*(1975)

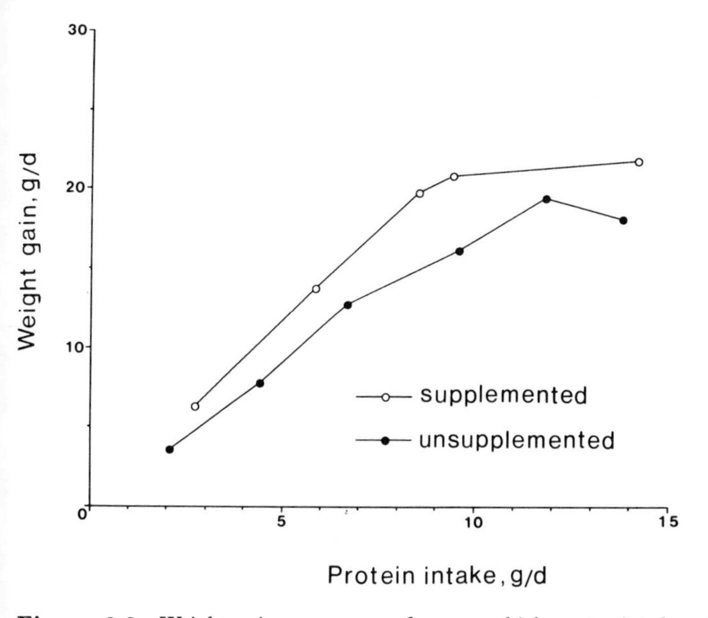

Figure 3.3. Weight-gain responses of young chickens to intake of protein from groundnut-wheat-barley mixtures unsupplemented or supplemented with lysine and methionine. Calculated from Wethli *et al.* (1975)

Table 3.1. RESPONSES OF YOUNG CHICKENS TO METHIONINE SUPPLEMENTED AND UNSUPPLEMENTED GROUNDNUT/LYSINE MIXTURES

Dietary protein (g/kg)	*Added methionine* (g/kg)	*Available methionine* (g/kg)	*Growth rate* (g/d)
200	-	2.1	5.8
200	2.1	4.2	9.5
240	-	2.6	7.9
240	1.6	4.2	11.1
320	-	3.5	10.6
320	0.6	4.1	10.9
370	-	4.0	10.6
370	0.5	4.5	10.8

(After Carpenter and De Muelenaere, 1965)
Growth rate on a well-balanced conventional control diet was 10.7g/bird d.

Wethli, Morris and Shresta (1975) compared the responses of young chickens to increasing concentrations of relatively poor-quality proteins, unsupplemented or supplemented with their limiting amino acids. In one experiment (Figure 3.2) a soyabean- maize mixture (Chemical Score about 70%) was fed unsupplemented or supplemented with methionine. The responses did not show the classical difference in slopes (c.f. Figure 3.1) and there was clear evidence that the maximum response to the unsupplemented protein was less than that to the supplemented protein. This difference (about 6%) was statistically significant. In another experiment a groundnut-wheat-barley mixture (Chemical Score about 47%) was supplemented with lysine and methionine in the adequate treatments (Figure 3.3). In this case, in respect of slopes, responses were more typical of expectations from theory, however there was again a poorer maximum response (about 13%) to the unsupplemented (poorer-quality) protein.

Wethli *et al* (1975) concluded from their experiments that 'maximum growth rate could not be obtained when groundnut meal or soyabean meal was used as a simple supplement to a cereal-based diet, even though very high dietary protein levels were used'.

Robinson and Boorman (unpublished) recognising some of the difficulties associated with sulphur amino acid deficiences and supplementations, mixed maize gluten meal (prairie meal) and soya protein isolate in fixed proportions (3:1, protein basis) and supplemented with lysine to produce three mixtures of different protein quality (poor, intermediate and good, Chemical Scores (lysine based): 62, 71 and 100% respectively). Although amino acids other than lysine were added to avoid obvious deficiences, it is probable that the most adequate mixture was marginally limiting in sulphur amino acids and arginine (Chemical Score about 92%).

These protein mixtures were fed at increasing concentrations in the diet to young chickens (Figure 3.4). The relationships among the slopes was as expected from the Chemical Scores and despite the evident difference in quality between good and intermediate mixtures, the maximum response achieved with the latter was equal to that achieved with the former. These two responses mimic theoretical expectations (Figure 3.1). However, there was evidence of a curtailed maximum response to the poor-quality protein mixture. This decrease (about 11%) was close to conventional significance ($P<0.10$).

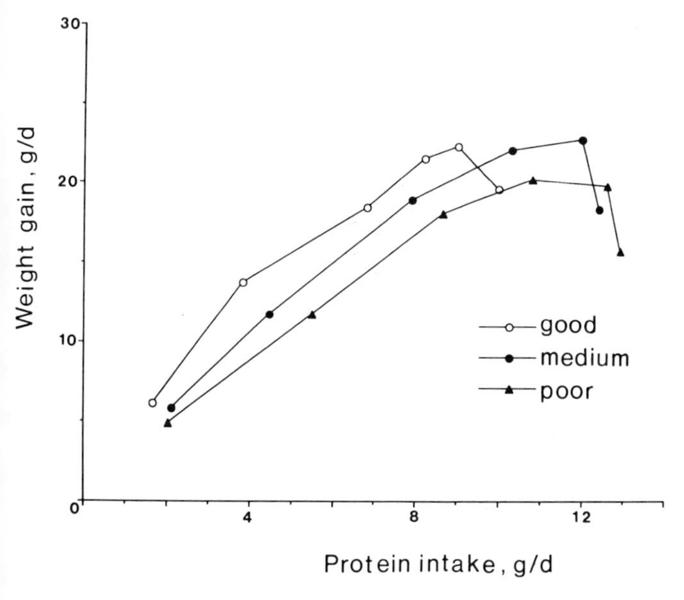

Figure 3.4. Weight-gain responses of young chickens to intake of protein from maize gluten-soyabean mixtures balanced with respect to other amino acids and unsupplemented or supplemented with lysine to provide three protein qualities (Robinson and Boorman, Unpublished)

It seems therefore that maximum response to protein, as defined by the response to good-quality protein, cannot always be obtained simply by feeding more of a poor-quality protein. It may be that for proteins of moderate quality (Chemical Score >70%) quantity can compensate for quality, but for poor-quality proteins this is not the case.

Utilisation of amino acids from poor-quality proteins

All proteins are mixtures of amino acids. Relative to the animal's requirements one essential amino acid is always likely to be in shorter supply than others. The response to protein is a response to this amino acid. It is relevant to ask whether there should be the expectation of poorer performance when this limiting amino acid is supplied from a poor-quality protein. That some should think that this is likely has arisen from studies on amino acid imbalance. This is a condition usually created by the addition of an incomplete mixture of essential amino acids to a balanced protein, thereby creating, in effect, a poor-quality protein mixture in which one (or more) amino acid is severely deficient. In comparisons between the responses to the balanced protein and to the poor-quality mixture, despite similar dietary concentrations of the limiting amino acid, responses to the latter were always substantially poorer than those to the former. Some attributed these adverse effects to impairment of the utilisation of the deficient amino acid in the presence of the relative excesses of other amino acids (Sauberlich and Salmon, 1955; Harper, 1959).

In the studies quoted above both Carpenter and de Muelenaere (1965) and Wethli *et al.* (1975) invoked amino acid imbalance as a likely cause of poorer performance from poor-quality proteins; the latter authors concluding that there was impaired utilisation of the most deficient amino acid in their studies.

The problem of impaired utilisation of the limiting amino acid in imbalanced and poor-quality protein mixtures was examined critically by Fisher, Griminger, Leveille and Shapiro (1960). They fed sesame meal as the sole source of protein in the diet of young chickens, varying the protein and lysine (limiting amino acid) concentrations and the degree of imbalance, by adding a mixture of amino acids lacking lysine (Table 3.2). Typical effects of imbalance were shown, i.e. reduced growth and food consumption, on the more imbalanced diets, especially at low protein concentrations. However, the authors examined intakes of lysine and concluded that for similar intakes of the limiting amino acid the responses in weight gain were essentially the same (see data indicated in Table 3.2). From such observations Fisher *et al.* (1960) concluded that imbalance among amino acids affects food intake primarily and that the growth depression is due to depressed intake of the limiting amino acid, not impaired utilisation. This can be illustrated more generally from the data in Table 3.2. If the three highest lysine intakes among the data are omitted as being on the plateau region of the curve (lysine not the limiting factor), a simple linear regression indicates that 98% of the variation in growth rate is accounted for by variation in lysine intake (Figure 3.5). The slope of this line represents nett utilisation of lysine for gain and indicates that in the deficient range, irrespective of protein concentration and quality (imbalance), utilisation is essentially constant and gain can be predicted from lysine intake.

Netke, Scott and Allee (1969) also examined this question, using crystalline amino acids and diets deficient in lysine, isoleucine or leucine. With respect to isoleucine (Table 3.3), one experiment tested the effect of increasing additions of an amino acid mixture lacking isoleucine on a balanced low-protein diet. Another experiment examined the responses to isoleucine in low and normal protein diets; the former case providing smaller excesses of other amino acids relative to isoleucine than the latter. These data represent several different states of protein concentration, isoleucine concentration and protein quality. In cases where isoleucine was in the limiting range (as identified by the authors), 96.5% of the variation in weight gain was accounted for by variation in isoleucine intake (current author's regression analysis, see Figure 3.6). The authors concluded: 'It would appear that excess amino acids, with a variety of patterns and concentrations, do not impair the utilisation of the first limiting amino acid even though their presence in the diet depresses feed intake and hence, weight gain.'

It is necessary to add one caveat to these clear conclusions. A few instances of specific amino acid antagonisms have been demonstrated in which an excess of one amino acid

Table 3.2. RESPONSES OF YOUNG CHICKENS TO BALANCED OR IMBALANCED DIETS AT DIFFERENT DIETARY CONCENTRATIONS OF SESAME PROTEIN AND LYSINE

Basal dietary protein[a] (g/kg)	*Imbalancing mixture added*[b] (g/kg)	*Dietary lysine content* (g/kg)	*Weight gain* (g/d)	*Food intake* (g/d)	*Lysine intake* (mg/d)
110	-	2.97	0.429	10.6	31.4
110	40	2.97	0.357	9.64	28.6
110	-	4.40	2.71	15.6	69.3
110	40	4.40	1.36	11.7	51.4[c]
110	-	7.70	10.6	29.9	231
110	40	7.70	10.0	24.7	191
140	-	3.78	1.36	12.5	47.9[c]
140	-	5.60	4.07	17.0	95.7
140	-	9.80	12.4	32.2	316
170	-	4.59	1.36	11.5	52.9[c]
170	-	11.9	16.4	32.1	381
200	-	5.40	2.43	13.6	74.3
200	-	14.0	18.1	33.3	477
230	-	6.21	2.29	12.1	75.0
230	40	6.21	2.57	14.0	87.1
230	-	9.20	11.3	28.1	259
230	40	9.20	9.64	24.9	229
230	-	16.1	19.4	34.5	555[d]
230	40	16.1	19.1	32.7	526[d]
260	-	7.02	3.64	15.6	109
260	-	10.4	13.2	30.5	317
260	-	18.2	20.4	35.4	644[d]

Adapted from Fisher *et al* (1960)

[a] Basal diets containing sesame meal as the sole source of protein and lysine, the limiting amino acid, was added to create different concentrations
[b] The imbalance was created by adding a mixture of amino acids lacking lysine
[c] Comparisons used in the text
[d] Data ommitted from regression analysis (see Figure 3.5)

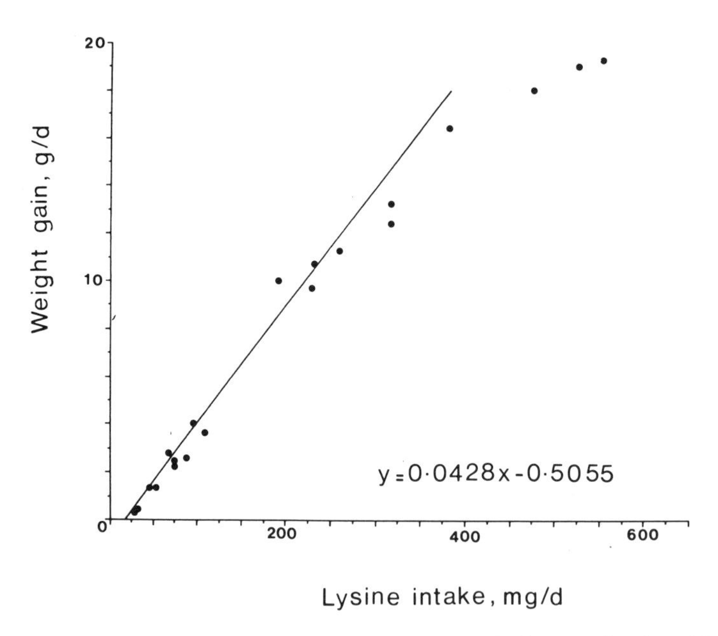

Figure 3.5. Weight-gain response of young chickens to intake of lysine from balanced and imbalanced sesame diets. Linear regression ($r^2 = 0.978$, $n = 19$) in the limiting range calculated from the data of Fisher *et al.* (1960) as shown in Table 3.2

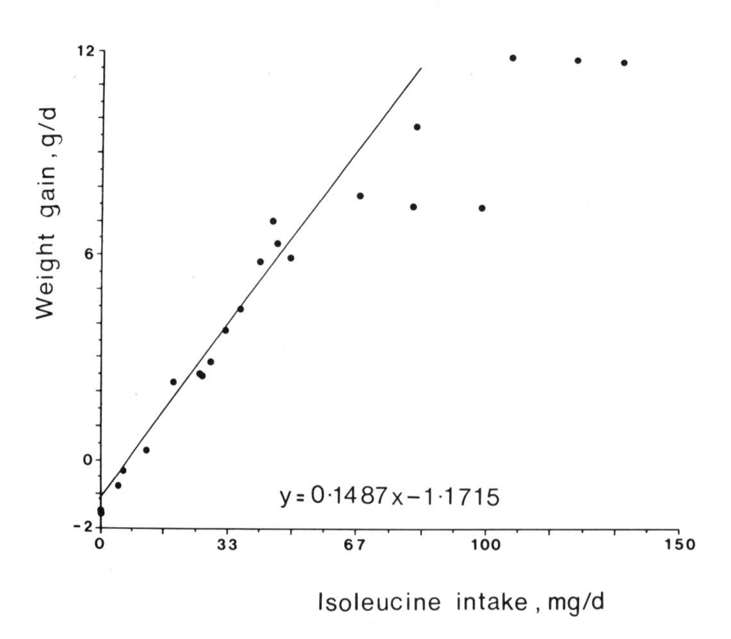

Figure 3.6. Weight-gain response of young chickens to intake of isoleucine from balanced or imbalanced synthetic amino acid diets. Linear regression ($r^2 = 0.965$, $n = 16$) in the limiting range calculated from data of Netke *et al.* (1969) as shown in Table 3.3

Table 3.3. RESPONSES OF YOUNG CHICKENS TO BALANCED OR IMBALANCED SYNTHETIC AMINO ACID DIETS AT DIFFERENT CONCENTRATIONS OF DIETARY NITROGEN AND ISOLEUCINE.

Dietary isoleucine (g/kg)	*Imbalancing mixture added*[a] (g/kg)	*Food intake* (g/d)	*Weight gain* (g/d)	*Isoleucine intake* (mg/d)
(Balanced low-nitrogen (15.4gN/kg) basal diet, Experiment 2)				
3	0	14.9	6.97	44.8
3	3.08	13.9	5.74	41.7
3	6.17	12.3	4.40	36.8
3	9.25	10.9	3.78	32.8
3	12.3	9.48	2.83	28.5
3	15.4	8.77	2.51	26.3
(Isoleucine-devoid, normal-nitrogen (30.8gN/kg) basal diet, Experiment 3)[b]				
0	-	3.65	-1.55	0
1	-	4.58	-0.78	4.67
2	-	5.93	0.28	11.8
3	-	8.87	2.46	26.7
4	-	12.4	5.91	49.7
5	-	16.4	9.75	81.8
6	-	17.8	11.8	107
7	-	17.6	11.7	124
8	-	17.0	11.2	136
(Isoleucine-devoid, low-nitrogen (18.5 gN/kg) basal diet, Experiment 3)[c]				
0	-	4.17	-1.47	0
1	-	5.75	-0.29	5.67
2	-	9.75	2.26	19.2
3	-	15.4	6.35	46.2
4	-	16.8	7.75	67.3
5	-	16.2	7.42	80.8
6	-	16.4	7.38	98.3

[a] The imbalance was created by adding a mixture of amino acids lacking isoleucine and is expressed here in nitrogen equivalents. Regression equations for weight gain (y) on isoleucine intake (x):
[b] Original authors for intakes of isoleucine $\leq$ 81.8 mg/d. $y = 0.139x - 1.374$
[c] Original authors for intakes of isoleucine $\leq$ 46.2 mg/d. $y = 0.168x - 1.265$
and, current author's for all points used by original authors plus all those for Experiment 2 ($n = 16$). $y = 0.149x - 1.172$ (see Figure 3.6)
Adapted from experiments 2 and 3 of Netke *et al* (1969)

interferes with the utilisation of another amino acid. Two cases, one in which lysine excess antagonises arginine and another in which the utilisation of one of the amino acids leucine, isoleucine or valine might be impaired by an excess of either or both of the other two, have been documented (Harper, Benevenga and Wohlheuter, 1970). It is possible for these to exist within the more general phenomenon of imbalance or poor quality. The data of Netke *et al.* (1969) (Table 3.3) probably provide an instance of this. In experiment 3 the isoleucine response at normal nitrogen was generated in diets including 12 g leucine and 8.2 g valine per kg, whereas the respective values at low nitrogen were 7.2 and 4.9 g. The original authors' regression analyses of these data did indicate a trend towards a lower response slope in the former case (see footnotes, Table 3.3), consistent with poorer nett utilisation of isoleucine for growth, although the authors attached no significance to this difference.

Within the range of excesses represented by poor-quality proteins, the effect of an antagonism may modify slightly the general response to excesses of amino acids. However, as noted above, cases are few and the dominant effect of such excesses on food intake is evident in the examples given above. This dominant effect and the lack of effect of general imbalance on the utilisation of the limiting amino acid was accepted by Harper *et al.* (1970) in their exhaustive and authoritative review of the subject. Much has also since come to be understood about the mechanism and significance of this effect on food intake (Boorman, 1979).

Responses to poor-quality proteins re-examined

Studies on amino acid imbalance were often performed with artificial diets and in protein-limiting circumstances. The latter condition seemed to exacerbate the responses observed. It is however reasonable to ask whether responses to relatively high intakes of intact poor-quality protein are consistent with those to imbalance. The resolution of this question requires re-examination of the responses shown in Figures 3.2, 3.3 and 3.4 in the form of responses to intakes of the respective limiting amino acids. Differences betweeen the slopes of the response relationships in the limiting range would be evidence of changes in limiting amino acid utilisation for growth.

Wethli *et al.* (1975) (Figure 3.2) identified methionine as the limiting amino acid in their soyabean-maize mixture. Re-expression of the responses on this basis (Figure 3.7a) is inconclusive, seeming to show poorer utilisation of methionine from the supplemented (good quality) source. This may have arisen from over-supplementation with methionine, or because part of the supplemented methionine must be used to form cysteine, a process which involves poorer efficiency of utilisation of methionine (Baker, 1977). Re-expression as responses to total sulphur amino acid intake produces similar slopes (Figure 3.7b) and offers no evidence of differences in sulphur amino acid utilisation. The conclusion from the experiment with ground-nut meal (Figure 3.3), where again methionine was identified as the limiting amino acid, is the same, and only the response to sulphur amino acids is shown here (Figure 3.8).

In choosing lysine as the limiting amino acid Robinson and Boorman (Figure 3.4) avoided some of the problems associated with sulphur amino acid utilisation. Their results, as responses to lysine intake (Figure 3.9), show no evidence of a difference in lysine utilisation between the poor and intermediate quality mixtures. Despite the possibility of a different limiting amino acid in the fully supplemented mixture (good) there

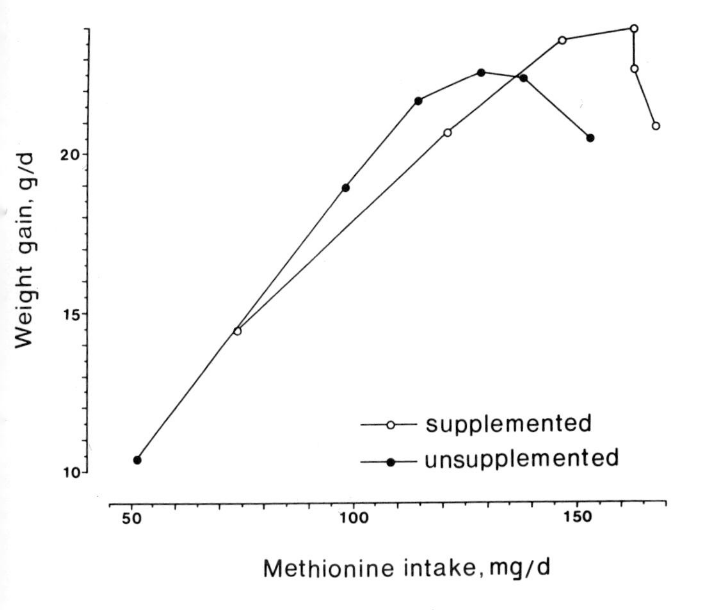

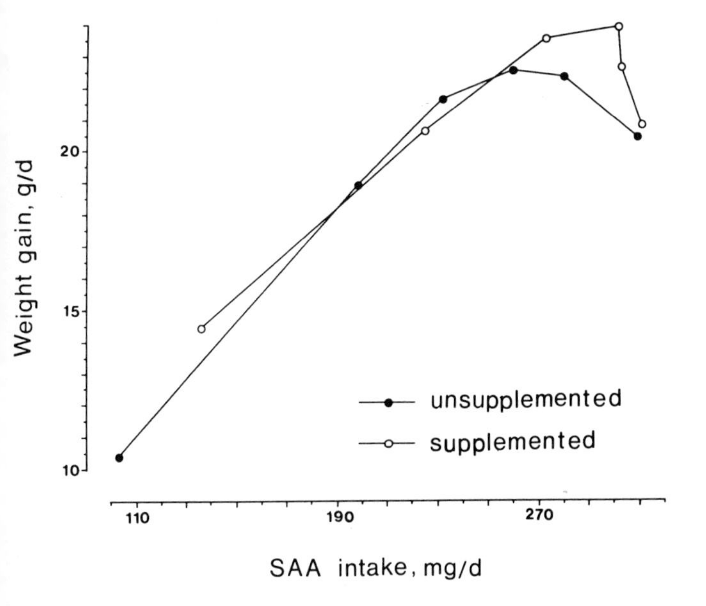

Figure 3.7. a. Weight-gain responses of young chickens to intake of methionine from soyabean-maize mixtures unsupplemented or supplemented with methionine. Calculated from Wethli *et al.* (1975), re-expression of Figure 3.2
b. Figure 3.7a re-expressed as responses to intake of sulphur amino acids (SAA)

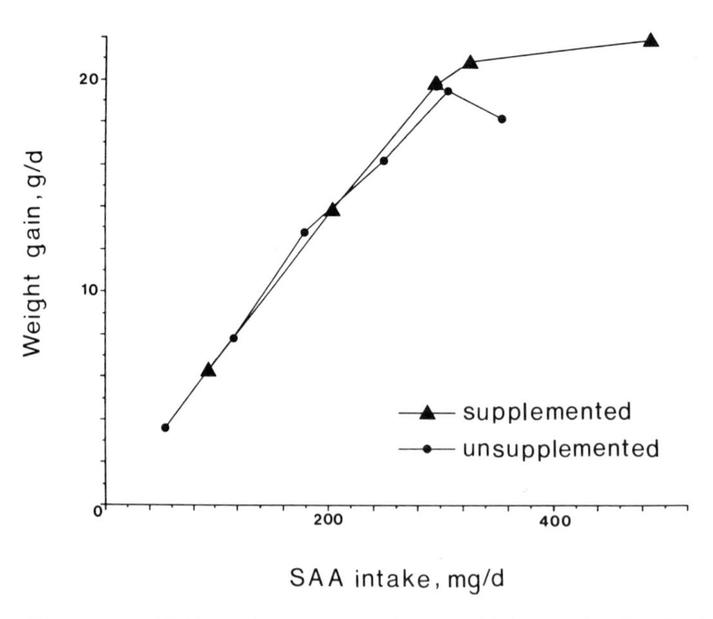

Figure 3.8. Weight-gain responses of young chickens to intake of sulphur amino acids (SAA) from groundnut-wheat-barley mixtures unsupplemented or supplemented with lysine and methionine. Calculated from Wethli *et al.* (1975), re-expression of Figure 3.3

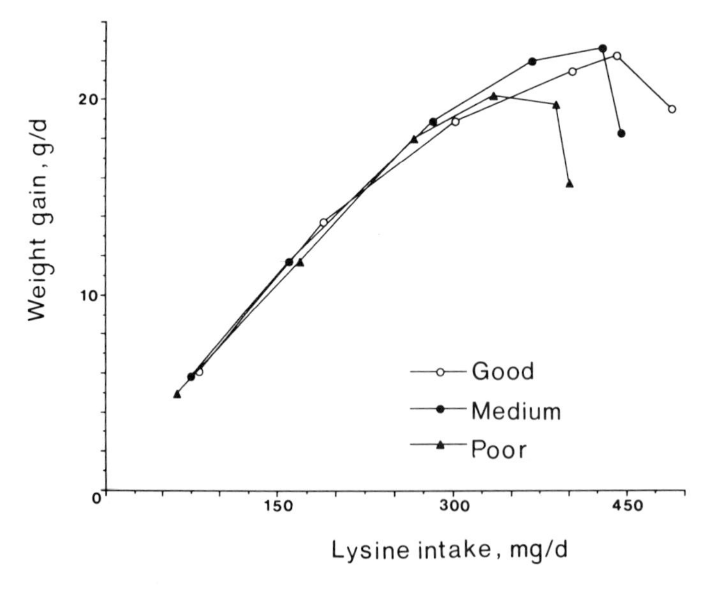

Figure 3.9. Weight-gain responses of young chickens to intake of lysine from maize gluten-soyabean mixtures balanced with respect to other amino acids and unsupplemented or supplemented with lysine to provide three protein qualities. Robinson and Boorman (Unpublished), re-expression of Figure 3.4

is also no evidence of a difference in utilisation in this case.

Other studies not directly concerned with responses to poor-quality proteins may be examined for evidence relating to utilisation of the limiting amino acid. Morris, Al-Azzawi, Gous and Simpson (1987) reported two trials (Reading and Natal) designed to examine the effect of protein concentration in the diet on the lysine requirement of the young chicken. They found that over the range 140 to 280 g protein/kg diet, the lysine requirement could be expressed as a constant proportion of the protein (about 54 g lysine/kg protein). These observations in their original form have been the subject of some discussion about lysine utilisation (D'Mello, 1988). This discussion cannot be repeated here but some points of particular relevance should be made. If the relationship between the protein concentration and requirement for an essential amino acid are consistent to the very high concentrations of protein which have been used in examining responses to poor-quality proteins, it is to be expected that the maximum response will be constrained, since by their nature poor-quality proteins will not contain the ideal concentration of the limiting amino acid. This is not a statement about mechanism, it is a restatement of the problem. Secondly, data such as those of Morris *et al.* (1987) can be represented as response curves to lysine in proteins of different quality (Figures 3.10a and b). The range of qualities is not large (Chemical Scores about 75 to >100), but in one trial (Figure 3.10a) the response to lysine in the poorest protein mixture (Chemical Score 77) does appear constrained throughout, although the rate of response is apparently not affected. Other responses appear indistinguishable from each other. In the other trial (Figure 3.10b) responses appear indistinguishable and offer no evidence of changes in utilisation of lysine.

The data of Morris *et al.* (1987) offer the rules for formulation at high protein concentrations, if resources are available to provide the ideal concentration of the limiting amino acid. This would not be a solution if poor-quality proteins had to be fed perforce without other resources. Abebe and Morris (1990) have extended the observations to tryptophan (12 g tryptophan/kg protein).

Within the limits of the results available therefore it seems that for the range in which the most deficient amino acid in the protein is also the growth limiting factor in the system, the conclusion of Fisher *et al.* (1960) about amino acid imbalance explains the response to poor-quality protein, i.e. if food intake is taken into account responses tend to become coincident in this region and there is no evidence of impaired utilisation of the limiting amino acid. Indeed poorer response to poor-quality proteins in this region falls within the general understanding of the response to protein (Figure 3.1). Other influences may contribute to a poorer response. In the experiments described, comparisons were made between a poor-quality intact protein and the same protein supplemented to good quality with free amino acids. Differences in amino acid digestibility may be expected in such comparisons. In some cases dietary concentrations of anti-nutritive factors may be increased beyond critical thresholds as poor-quality proteins are increased in the diet. Digestibility is discussed further below and effects of anti-nutritive factors would probably be recognised in other ways, if they were impairing response.

In the plateau region of the response curve, the most deficient amino acid in the dietary protein is no longer the limiting factor in the system. The slope of the response line tends to zero and comparison between good- and poor-quality protein sources in terms of utilisation of the limiting amino acid has little meaning. Neither can the difference in maximum response be explained in terms of intake of the limiting amino acid

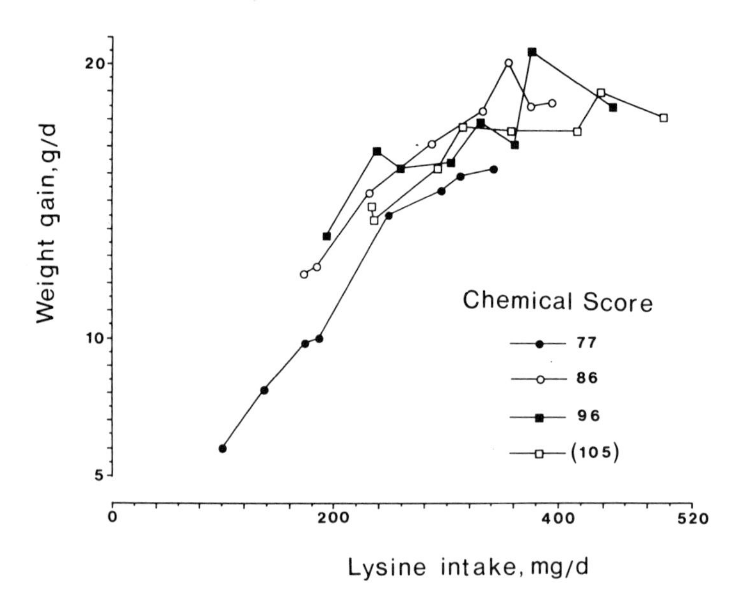

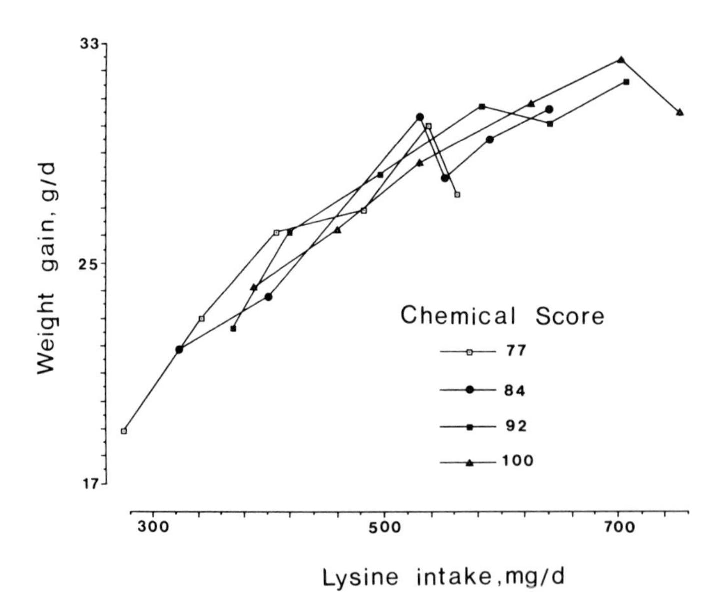

Figure 3.10. Weight-gain responses of young chickens to intake of lysine from protein mixtures of different quality. Calculated from Morris *et al.* (1987). Chemical scores relative to requirement pattern (Boorman and Burgess, 1986) calculated by current author.
a Trial 1/ (Reading University),
b Trial 2/ (Natal University)

since growth differences are seen at equal intakes of limiting amino acid (Figures 3.7, 3.8 and 3.9). It is, then, in respect of maximum response that explanation of poorer growth rate from poorer-quality proteins requires explanation and it is in this region of the response curve that theory and observation are not always in accord.

Two points of equal intake of the limiting amino acid from poor and good-quality protein mixtures differ in that the former intake is provided from a diet much higher in protein than is the latter. There has been little study of the animal's physiological and biochemical responses to high intakes of imbalanced protein. Wethli *et al.* (1975) noted that dietary concentrations of balanced protein equivalent to those of imbalanced protein did not cause the growth impairment. The cause of the phenomenon does therefore seem to reside in the imbalanced nature itself. Possibly the mechanism is related to changes in the relationship between protein and the non-protein energy contents of the diet. It is also known that at lower protein intakes imbalanced protein causes raiding of the animal's tissue (muscle) for the limiting amino acid (Harper *et al.* 1970). If this occurs at high intakes of imbalanced protein it might provide sufficient explanation of the phenomenon. In the absence of detailed information more speculation is unwarranted. Robinson and Boorman (unpublished) observed that responses in nitrogen retention were similar to those in body-weight gain (Figure 3.9), so accumulation of fat would not explain the difference.

Whatever the mechanism involved the effect, when observed, is relatively small and produced by extreme protein concentrations. Carpenter and de Muelenaere (1965) used diets up to 370 g protein/kg, Wethli *et al.* (1975) 420 (groundnut) and 300 g protein/kg (soyabean) and Robinson and Boorman 560 g protein/kg. Impairment of maximum response would not be encountered in the conventional substitution of protein for synthetic amino acids which might occur in normal commercial practice currently. The impairment would also be of insufficient magnitude in most circumstances to be of importance in situations where poor-quality proteins would be used perforce.

Minimising dietary protein concentration

It follows from the foregoing that supplying a sufficiency of a limiting amino acid from a poor-quality protein leads to a high-protein diet containing large excesses of amino acids other than the limiting one. These excesses cannot be utilised and must ultimately be catabolised, leading to increased excretion of nitrogen largely as uric acid. A more balanced protein which can supply an adequacy of the limiting amino acid at a lower dietary concentration without large excesses of other amino acids must therefore lead to more efficient utilisation of dietary protein and less nitrogen excretion. This is becoming increasingly important in relation to litter quality and its effect on bird welfare and carcass quality, as well as to nitrogen disposal from poultry enterprises.

Protein quality can be optimised by judicious use of good-quality protein sources, mixing complementary sources of protein and using synthetic amino acids. Synthetic amino acids allow addition of the limiting amino acid without increasing concentrations of non-limiting amino acids, i.e. without increasing protein concentration. In this sense synthetic amino acids are a substitute for protein. Over a limited range this procedure is used routinely in ration formulation. The use of DL-methionine allows a cereal-soyabean diet to meet a young chicken's requirement for sulphur amino acids at about 240 g protein/kg, whereas supplying the sulphur amino acids from intact protein requires a diet of about 270 g protein/kg. Replacement of other amino acids in the limiting sequence allows successive reductions in dietary protein concentration. There is a belief that an increasing array of synthetic amino acids will become commercially available. This possibility raises two questions:

1. Is there a unique feature of intact protein which cannot be provided by synthetic amino acids, and

2. Is limitation of non-essential nitrogen likely to impose a constraint?

Synthetic amino acids

It is usually assumed that synthetic amino acids are freely and completely absorbed from the intestine, whereas intact proteins are unlikely to be completely digested. This suggests that synthetic forms offer a more effective supply of amino acids. However substantial proportions of the amino acids from intact dietary proteins are absorbed as small peptides and this may represent the optimal form for the processes in the gut mucosal cell. It is also possible that free amino acids are more susceptible to bacterial degradation, because of the greater proportion of unreacted amino and carboxyl groups in this form. The extent of bacterial activity in the small intestine should not be underestimated. The idea that synthetic amino acids are a more effective source of supply from the intestine should not therefore be accepted without question. Post-absorptive utilisation may also be impaired. Batterham (1980) found very poor utilisation of synthetic lysine (50 to 60%) in pigs fed once daily. This was attributed to different rates of delivery of the free lysine and the amino acids from intact protein to sites of synthesis. Lighting patterns of poultry, which largely dictate feeding patterns, might therefore be of importance in determining utilisation of synthetic amino acids.

There is a wealth of evidence in poultry relating to comparison of intact proteins with synthetic amino acids. This arises from the fact that synthetic amino acids are used as reference forms (standards) in biological assays for assessing amino acid availability in intact protein sources. If synthetic amino acids were more poorly utilised than amino acids from protein sources, it would be expected that availabilities of amino acids from well digested proteins would frequently be found to be well in excess of 100%. In practice such values are very unusual. It may be concluded therefore that for chickens fed *ad libitum* there is no reason to suppose that synthetic amino acids are poorly utilised relative to amino acids from intact protein. This also accords with experience of using synthetic amino acids in commercial formulation.

Non-essential nitrogen

In addition to essential amino acids protein provides 'non-essential nitrogen'. It is important to realise that this not only arises from non-essential amino acids (e.g. glutamic acid, aspartic acid, alanine), but will also arise from essential amino acids provided in excess of their requirements (Bedford and Summers, 1985). Furthermore, estimates of nett efficiency of utilisation of absorbed essential amino acids for growth are about 60 to 80% (Boorman and Burgess, 1986), so that inefficiencies of utilisation will provide another source of nitrogen to the non-essential pool. Efficiencies of supply of nitrogen to the non-essential pool from different essential amino acids may differ (Heger, 1990).

In the most complete study of the subject with chickens, Bedford and Summers (1985) fed diets containing balanced patterns of essential amino acids (EAA) at 0.35, 0.45, 0.55 or 0.65 of dietary protein concentrations of 140, 180 or 220 g/kg. They found that response was maximised at an intake of 5 g EAA/day, that 0.55 EAA per unit protein was about optimal and that the response to protein increased throughout the range used. It may be estimated from these results that a diet of about 180 g protein/kg containing a balanced pattern of EAA to provide an intake of about 5 g EAA per day would represent a minimum state of adequacy. This diet would include about 80 g non-essential amino acid nitrogen (expressed as protein) per kg. It should be noted that this refers to EAA minima in a balanced pattern and therefore includes little or no contribution from EAA in excess of their requirements.

Replacement of protein by synthetic amino acids

Recent concerns have stimulated several studies of the efficacy of replacing intact protein with synthetic amino acids, although it is a subject that has been researched intermittently for many years. The form that these studies usually takes is the comparison of a diet conventionally-formulated to meet the amino acid requirements of the chicken, as prescribed by a recognised authority, with a similar diet or diets in which protein is reduced and amino acid requirements are maintained by additions of synthetic amino acids. It should be noted that the conventional control diet is usually formulated to contain DL-methionine, thus implying partial acceptance of the principle which is being tested.

Table 3.4. RESPONSES OF YOUNG CHICKENS TO SUCCESSIVE REDUCTIONS IN DIETARY PROTEIN COMPENSATED BY SUPPLEMENTATION WITH SYNTHETIC AMINO ACIDS TO MAINTAIN REQUIREMENT MINIMA.

Dietary protein (g/kg)	*Added amino acids*	*Food intake* (g/d)	*Weight gain* (g/d)
230	M	50.9	36.4
200	MLATh	49.4	33.8
170	MLAThTpVI	50.0	31.8

M: methionine, L: lysine, A: arginine, Th: threonine,
Tp: tryptophan, V: valine, I: isoleucine
(Selected data from Pinchasov *et al.* 1990)

Edmonds, Parsons and Baker (1985) found that performance of young chickens was not maintained when protein was decreased from 245 g to 160 g/kg diet with appropriate supplementation. In the latest of a series of studies (see also Fancher and Jensen, 1989), Pinchasov, Mendonica and Jensen (1990) compared a conventional diet (230 g protein/kg) with diets containing 200 g or 170 g/kg appropriately supplemented (see Table 3.4 for selected data). They found a decline in performance of young chickens about commensurate with the decrease in dietary protein. Holsheimer and Janssen (1991) came to similar conclusions from experiments with broilers in the grower/finisher period (3 to 7 weeks) using diets in which protein was reduced to 170 g protein/kg. In this work there were some differences between sexes and it may be relevant that a repeating 1 h light: 3 h dark lighting pattern was used. Experiments with starting chickens seem usually to have involved continuous light.

Table 3.5. RESPONSE OF YOUNG CHICKENS TO SUCCESSIVE REDUCTIONS IN DIETARY PROTEIN COMPENSATED BY SUPPLEMENTATION WITH SYNTHETIC AMINO ACIDS TO MAINTAIN REQUIREMENT MINIMA

	Experiment 4		*Experiment 5*		
Dietary protein (g/kg)	230	178	230	174	169
Added amino acids	M	MLATh TpG	M	MLATh TpIPG	MLATh TpIPLeG
Food intake (g/d)	56.1	61.6	50.5	56.0	54.3
Weight gain (g/d)	36.7	38.0	33.0	34.6	33.5

(Selected data from Parr and Summers, 1991) M: methionine, L: lysine, A: arginine, Th: threonine, Tp: tryptophan, I: isoleucine,
P: phenylalamine, Le: leucine, G: glycine

The results described are the more puzzling because other similar studies have shown that compensation for reduced protein by additions of synthetic amino acids is possible. Table 3.5 shows selected data from a large series of experiments by Parr and Summers (1991). These authors consistently found that growth on low-protein diets could be restored by appropriate supplements of synthetic amino acids. Two points about these studies are worthy of mention — they found that the tryptophan requirement of the National Research Council (1984) was not adequate for full restoration of growth and in several cases supplemented low-protein diets produced better performance than the conventional control diet. The former point should be borne in mind by others using National Research Council (1984) as the standard in such comparisons and the latter may indicate an inadequacy in the control diet. In studies reported in abstract Stilborn and Waldroup (1988, 1989) showed equivalent performance in broilers (3 to 6 weeks) fed on conventional diets and low-protein diets supplemented with synthetic amino acids.

Pinchasov *et al.* (1990) discussed their finding that synthetic amino acids fail to compensate for protein in terms of some of the possible mechanisms described above. They also tested for inadequacy of non-essential nitrogen, using glutamic acid addition and were able to reject this as a cause of poorer performance. This is perhaps not surprising if excess essential amino acids are also a source of non-essential amino acid nitrogen. Parr and Summers (1991) also showed some substantial reductions in dietary protein that did not cause deficiencies of non-essential nitrogen.

Thus, a state of confusion exists in relation to the use of substantial proportions of synthetic amino acids. If there is a nutritional difference between synthetic forms and protein, it remains to be explained why the former seem to be adequate standards for bioassays, why experience over a limited range with DL-methionine and L-lysine has given no indication of a difference in performance and why some experimenters find no evidence of such a difference. If there is no nutritional difference between the two forms then the consistency with which some experimenters find evidence of a difference in performance is a matter of genuine concern. Much effort has been expended in deriving data relating to lysine and methionine. The quantity and quality of the information relating to other amino acids is generally poorer. It may simply be in the reliability of requirement estimates, ingredient compositions and aspects such as availability that the source of much of the confusion lies. However, until there is a clear resolution of the question the use of more synthetic amino acids as replacements for protein will continue to be regarded with caution.

References

Abebe, S. and Morris, T.R. (1990). *British Poultry Science*, **31**, 267–272

Allison, J.B. (1964). In *Mammalian Protein Metabolism, volume 2*, pp 41–86. Edited by H.N. Munro and J.B. Allison. New York: Academic Press

Baker, D.H. (1977). *Advances in Nutritional Research*, **1**, 299–335

Batterham, E.S. (1980). In *Recent Advances in Animal Nutrition - 1979*, pp 11–22. Edited by W. Haresign and D. Lewis. London: Butterworths

Bedford, M.R. and Summers, J.D. (1985). *British Poultry Science*, **26**, 483–491

Boorman, K.N. (1979). In *Food Intake Regulation in Poultry*, pp 87–126. Edited by K.N. Boorman and B.M. Freeman. Edinburgh: British Poultry Science

Boorman, K.N. and Burgess, A.D. (1986). In *Nutrient Requirements of Poultry and Nutritional Research*, pp 99–123 Edited by C. Fisher and K.N. Boorman. London: Butterworths

Carpenter, K.J. and de Muelenaere, H.J.H. (1965). *Proceedings of the Nutrition Society*, **24**, 202–209

D'Mello, J.P.F. (1988). *World's Poultry Science Journal*, **44**, 92–102

Edmonds, M.S., Parsons, C.M. and Baker, D.H. (1985). *Poultry Science*, **64**, 1519–1526

Fancher, B.I. and Jensen, L.S. (1989). *Poultry Science*, **68**, 1385–1395

Fisher, H., Griminger, P., Leveille, G.A. and Shapiro, R. (1960). *Journal of Nutrition*, **71**, 213–220

Harper, A.E. (1959). *Journal of Nutrition*, **68**, 405–418

Harper, A.E., Benevenga, N.J. and Wohlheuter, R.M. (1970). *Physiological Reviews*, **50**, 423–558

Heger, J. (1990). *British Journal of Nutrition*, **64**, 653–661

Holsheimer, J.P. and Janssen, W.M.M.A. (1991). *British Poultry Science*, **32**, 151–158

Morris, T.R., Al-Azzawi, K., Gous, R.M. and Simpson, G.L. (1987). *British Poultry Science*, **28**, 185–195

National Research Council (1984). *Nutrient Requirements of Poultry, 8th revised edition.* Washington DC: National Academy Press

Netke, S.P., Scott, H.M. and Allee, G.L. (1969). *Journal of Nutrition*, **99**, 75–81

Parr, J.F. and Summers, J.D. (1991). *Poultry Science*, **70**, 1540–1549

Pinchasov, Y., Mendonica, C.X. and Jensen, L.S. (1990). *Poultry Science*, bf 69, 1950–1955

Sauberlich, H.E. and Salmon, W.D. (1955). *Journal of Biological Chemistry*, **214**, 463–473

Stilborn, H.L. and Waldroup, P.W. (1988). *Poultry Science*, **67** (Supplement 1), 36 (Abstract)

Stilborn, H.L. and Waldroup, P.W. (1989). *Poultry Science*, **68** (Supplement 1), 142 (Abstract)

Wethli, E., Morris, T.R. and Shresta, T.P. (1975). *British Journal of Nutrition*, **34**, 363–373.

II

Ruminant Nutrition

4

NUTRITION - GENOTYPE INTERACTIONS IN DAIRY CATTLE

J. D. OLDHAM[1], G. SIMM[1] and S. MARSDEN[2]

[1] *SAC Edinburgh, Bush Estate, Penicuk, EH26 0QE*

[2] *Dalgety Agriculture Ltd, 180 Aztec West, Almondsbury, Bristol BS12 4QH*

Introduction

In his introduction to the Proceedings of the Symposium on the 'Effect of genetic variance on nutritional requirements of animals', Sunde (1975) quotes Schneider (1946) as saying "that scientists interested in nutrition or in genetics might well be in different worlds". He goes on to say that, the separation of interest between nutritionists and geneticists had not altered in the intervening years from 1946 to 1974, when that symposium was held.

Even since 1975 it would seem that, amongst those interested in ruminants at least, this unfortunate trend of a separation in outlook between the nutritionist and the geneticist has continued with developments in nutrition largely being generated with little attention to genetics and *vice versa.* This is much less true for the non-ruminant sector (pigs and poultry) where intense selection has yielded remarkable rates of genetic improvement over the last 2 decades and the nutritionist has been forced to consider the implications of possible genotype x nutrition interactions.

Compared with non-ruminants rates of progress in genetic improvement with dairy cattle have, because of long generation intervals, necessarily been slower - although in North America annual rates of improvement of 1.2 - 1.4 per cent for milk, milk fat and milk protein are currently being achieved (Lawlor and Short, 1990) - these rates being about half maximal (Woolliams, 1990) for conventional progeny test evaluation (Table 4.1).

Until quite recently, rates of genetic improvement in the UK have lagged quite far behind those in other countries. For example, between 1975 and 1985 the rate of improvement (estimated from average merit MMB sires used (MMB, 1985)) for fat and for protein yields respectively, were around 0.35 and 0.3 per cent per year. This represented 20–25% of the overall improvement in yield of milk solids over this period. More recently, over the years 1985 to 1990, Coffey (1991) has used records of pedigree indexes to calculate rates of genetic improvement in the indexed population in the UK and has identified annual rates of genetic improvement in milk, fat and protein yield of 1.6–1.8 per cent per year. While it could not be expected that such rates would apply to

Table 4.1. THEORETICAL RATES OF GENETIC IMPROVEMENT IN OUTPUT PERFORMANCE OF DAIRY COWS USING DIFFERENT BREEDING TECHNIQUES (FROM WOOLLIAMS, 1990)

System	*Rate of progress* (%/year)
Conventional progeny test	1.5 - 2
Efficient progeny test	2.6 - 3.3
Adult MOET	2.4 - 3.1
Juvenile MOET	3.4 - 4.5

MOET = Multiple Ovulation and Embryo Transfer

the UK population as a whole, the magnitude of the values shows a considerable shift in attention to genetic improvement in a significant section of the national dairy herd. These improvements build on the long-established technology of artificial insemination (AI) but perhaps reflect a growing awareness of the reliability of bull proofs for performance and the value of genetic improvement towards performance goals. Perhaps also the widespread availability of Cow Genetic Indexes (CGIs) and the excitement which has been aroused by the current, or pending, availability of new breeding technologies (perhaps particularly embryo transfer) has raised the issue of objective breeding more forcefully than in the past. It seems reasonable to suggest that availability of these new technologies can, from now, be expected to enable an enhanced rate of genetic progress into the foreseeable future (see Woolliams, 1990 and Table 4.1).

It is coincidental that these techniques have come to, or near to, the point of application at a time when market pressures on the dairy industry have changed and, indeed, have increased as the supply of milk products has outstripped demand. With quota limits on output, naturally attention has turned to the control of inputs for the herd - although this does not deny a place for continued high input, high output for individual cows in some circumstances. The intriguing counterpoint of enhanced capability for achieving genetic progress, together with tighter control on food input does, however, raise the issue of common ground between the nutritionist and the geneticist into higher profile than has needed to be the case up to now.

In light of these changes in availability of methodologies to achieve genetic progress and the shifting demands of the market, one important question concerns the choice of future selection goal(s). With feed costs contributing a high proportion of overall input costs for the dairy cow, interest in efficient feed use has been reawakened in the last few years (Korver, 1988; Gibson, 1986; Simm *et al.*, 1991; Oldenbroek, 1986) and associations between feed efficiency and profitability are actively being addressed.

So a number of developments call for an appraisal of nutrition/genetic interactions. First, where low input systems of feeding are practised, do selection criteria differ from the high input/high output situation? Second, should food intake and/or efficiency of feed use (feed efficiency) figure in the development of future selection indexes? Third, does the acceleration in genetic progress now being achieved call for adjustment to feeding strategies? The first of these issues calls for consideration of possible genotype x environment interactions.

Genotype × environment interactions

Genotype x environment interactions (GxE) obviously encompass factors of management, the physical environment, the thermal environment and others, as well as just feeding and nutrition. However, genotype x nutrition (GxN) interactions are often considered as a major part of GxE interaction and the majority of the literature from temperate regions on GxE is fairly specifically related to GxN.

Freeman (1967 and 1975) has reviewed much of the earlier information on GxE interactions in dairy cattle. For production (yield of milk or milk solids) Freeman's (1975) general conclusion was that, whilst there is some evidence for GxE interaction, the magnitude of the interaction is generally small and not sufficient to give concern for making selection decisions in the range of environments likely to be expected in the prevailing commercial populations. The current applicability of that observation might be challenged on the grounds that both phenotypic performance and genotype have been improved considerably from some of the earlier studies. In particular there might be reason to think that the genetic progress which has been achieved over the last two decades in yield capacity of dairy cows, has begun to outstrip the capacity of cows to eat sufficient to sustain those higher yields and to remain in good health on all feeding systems which might be used in practice. Feed systems which emphasise use of forage might cause GxN interactions which previously were small to take an exaggerated importance with some of today's genotypes. Certainly the evidence from North America (see Pearson *et al.*, 1990) would suggest that genetic progress has been making an ever-increasing proportionate contribution to realised output per cow in recent years and is now making a larger contribution to overall improvement than are other management factors. This trend can be expected to show also in the UK. Interactions can be considered either between or within breeds.

GxE-Between breed studies

In comparisons between Jersey cattle and other dairy breeds, interactions between genotype and feed environment (GxN) have been found (Oldenbroek, 1986a, b; Table 4.2). The differences applied both to feed intake and milk production traits, with the difference between Jerseys and other breeds being less on a high roughage diet than on a high concentrate diet. In large part, variations in relative performance were attributed to the higher relative feed capacity of Jerseys compared with other breeds. This between breed difference in intake capacity has been noted previously (Gibson 1986; Brigstocke *et al.*, 1982). In the Oldenbroek (1988) data, scaling of liveweight to power 0.58 (instead of 0.75) would remove the between breed difference seen with high concentrate feeding so it could be argued that the difference between Jersey cattle and others in feed intake is merely a question of scaling. But the difference in intake would remain for high roughage feeding even with the alternative weight scale. It would therefore seem that the difference is real and perhaps accentuated with higher forage feed systems.

Table 4.2. EVIDENCE OF BETWEEN BREED INTERACTIONS WITH FEEDING ENVIRONMENT. DATA ARE FOR COMPARISONS OF PERFORMANCE OF JERSEY HEIFERS WITH A MIXED SAMPLE OF HOLSTEIN-FRIESIAN, DUTCH FRIESIAN AND DUTCH RED AND WHITE CATTLE('LARGE DAIRY') AND ARE TAKEN FROM OLDENBROEK, (1986)

Breed	*Jersey*		*Large Dairy*		GxE
Diet (%Conc)	0	50	0	50	
DMI (kg/d)	10.5	12.7	13.1	16.3	$P<0.05$
Yield: Milk (kg)	10.1	11.6	13.7	18.1	$P<0.05$
Fat (g)	626	758	601	777	NS
Protein (g)	385	480	443	623	$P<0.05$
Efficiency (milk MJ/MJ ME)	0.43	0.39	0.37	0.36	$P<0.05$

Taylor, Moore and Thiessen (1986b) in comparisons amongst 25 British breeds of cattle also identified Jerseys (as well as Beef Shorthorns) as having particularly high scaled rates of food intake, particularly at later ages which would obviously apply to lactating animals. They indicated that the log-log regression between food intake and body weight yielded a regression coefficient which fell with age from around 0.85 at 5 months to around 0.7 approaching 3 years of age, identifying the need for care in scaling appropriately when considering between breed differences in intake. Only Oldenbroek (1988a, b) has made comparisons between breeds in different feed systems and his data point to an important GxE interaction which, in particular, raises issues about the capacity of different animals (in his case, different breeds) to ingest foods of lower nutrient density.

Within breed

GxE interaction can be of two kinds - either a different ranking of genotypes in different environments, or altered differences between genotypes in different environments.

There are few examples of the first kind - although some results from a substantial study in which top proven AI sires from Canada and New Zealand evaluated in both the Canadian and New Zealand environments (Burnside *et al.*, 1990; Peterson, 1988) have shown a re-ranking of Canadian sires according to the performance of their progeny in New Zealand, compared with expectations from evaluations in Canada.

Twenty sires, proven in their country of origin were used and their progeny's performance evaluated in 20 New Zealand herds and 10 Canadian herds. In each country, progeny of Canadian sires yielded more milk and milk solids than the New Zealand sire group - but concentrations of both fat and protein were higher for the progeny of NZ sires (Table 4.3).

Table 4.3. PERFORMANCE OF PROGENY OF CANADIAN OR NEW ZEALAND SIRES EVALUATED EITHER IN CANADA OR IN NEW ZEALAND

Trait	*Location (for evaluation):*	*Sire Group* Canadian	*New Zealand*
Milk (kg)	Canada[1]	6097	5469
	New Zealand[2]	3395	3157
Fat (kg)	Canada[1]	231	226
	New Zealand[2]	139.7	137.9
Protein (kg)	Canada[1]	206	192
	New Zealand[2 3]	108.7	104.6
Fat (g/kg)	Canada[1]	NR	NR
	New Zealand[2]	40.7	43.3
Protein (g/kg)	Canada[1]	NR	NR
	New Zealand[2]	31.8	33.1

[1] Data from Burnside *et al.* (1990)
[2] Data from Peterson (1988)
[3] Calculated from fat, lactose and energy yield data of Burnside *et al.* (1990)

The correlations between CANZ trial proofs and original Canadian proofs (Table 4.4) show that, for yield traits, but not for composition, the correlations for Canadian sires were significantly less than expected, indicating a major GxE interaction such that the ranking of sires (as the performance of their progeny) was not the same for Canadian bulls evaluated in New Zealand as would be expected from the initial progeny evaluation in Canada.

The most obvious environmental difference between Canadian and New Zealand management systems is the emphasis on pasture in New Zealand compared with the high level of grain feeding in Canada. The results of this trial therefore raise important questions about the possibility of differences in food, especially forage, intake capacity amongst cattle which might lead to certain individuals having a competitive advantage in lower input feeding regimes, which would not be apparent in feed systems which rely more on high nutrient density systems.

Table 4.4. CORRELATION (r) BETWEEN CANZ TRIAL AND CANADIAN SIRE PROOFS FOR PRODUCTION

Trait	*Canadian Sires*	*New Zealand Sires*
Milk (kg)	0.22[1]	0.80
Fat (kg)	0.25[1]	0.79
Protein (kg)	0.36[1]	0.80
Fat (g/kg)	0.69	0.75
Protein (g/kg)	0.68	0.54
Expected correlation	0.68	0.65

[1] Significantly different from expected
Data from Peterson (1988)

Apart from this study, direct evidence for within breed variation in food intake capacity, which is sufficient to give pause for thought in making selection decisions for particular environments is not so clear.

In our own studies of high and average genetic index (for production) dairy cattle offered diets containing, on average, 500 or 750g silage DM/kg diet DM (see Simm *et al.*, 1991 and below for details) it is noticeable that the coefficient of variation (CV) in DM intake amongst similar genotypes offered a lower proportion of forage in the diet is somewhat less, at 9.4% than with the higher forage diet (CV = 12.6%). CV for milk yield in the two feeding systems, however, is much the same (CV = 14.7 - 12.9% respectively). Korver (1982) recorded *ad libitum* intake of hay (experiment 1) or silage (experiment 2) by Holstein- Friesian or Dutch Friesian cows offered fixed allocations of concentrates. The intake of concentrate and silage corresponded to diets containing about 130 or 450g concentrate DM/kg total DM and in that study CV of forage intake, at 3.7 - 10.0% was not different for the two concentrate allowances, but was also lower than found in our own studies at Langhill. Whether this difference between the two sets of observations is meaningful is uncertain. The average DOMD of the hay and silage used by Korver (1982), at around 0.65 was, if anything, slightly lower than would apply for the silages used at Langhill so the possibility of differences in forage quality being a cause of the difference between the two sets of observations seems to be ruled out.

In pasture-based systems, cows of high genetic merit for milk production did not have higher *ad libitum* pasture intakes than their lower merit contemporaries when comparisons were made for Friesian cattle (Grainger *et al.*, 1985a) although in comparison of high and low breeding index Jersey cattle (Bryant, 1986) high breeding index animals ate around 14 per cent more pasture than low index animals. From these studies, though, there is some confounding with size as, in Jerseys, high index animals were up to 40 kg (around 10%) heavier than low index animals (Bryant, 1986) whereas in the Friesian study (Grainger *et al.*, 1985a) low index animals were slightly heavier than high index ones.

Even where intake is similar between individuals there may be differences in digestive function which could possibly lead to different amounts of nutrients being extracted from the food consumed. In general it has been assumed (eg. Freeman, 1975) that differences in digestive functions between animals are quite small. But Orskov *et al.* (1988) have raised the intriguing prospect of there being consistent differences between individual cows in the rate of passage of food particles through the gut. The variability they observed in fractional rate of particulate passage (cv. approx 20%) was large compared with the variability in estimates of digestibility (cv. less than 5%) and also larger than the estimates of CV for food and forage intake indicated above. Whilst some of this difference probably reflects the difficulties of achieving the same technical accuracy in measures of particle passage, compared with digestibility or intake, there may be grounds to consider that, across different kinds of environment, animals with inherently different digesta processing characteristics may have different advantages. Evaluation by modelling of digestive function across ruminants (and indeed other herbivores) of different size would certainly show this to be an important possibility (Illius and Gordon, 1991).

Food Intake, Feed efficiency and Profitability - The Langhill Experience

Until recently there have been few data with which to investigate the consequences of genetic improvement on feed efficiency, or to explore possible GxN interactions under UK feeding conditions, either intensive or extensive. Gibson's (1986) data on feed efficiency and performance of genetically high and low milk producing British Friesian and Jersey cattle deal only with a single complete pelleted diet offered *ad libitum*. In both breeds high potential production progeny produced more liquid milk, fat and protein, but at a lower fat and protein concentration than their low potential contemporaries. High potential progeny of both breeds had higher food conversion efficiencies to liquid milk, fat and protein than low potential production progeny. Food intake in Jerseys, scaled to weight, was greater than that in Friesians.

These data are useful in pointing out the association between selection for performance and feed efficiency but their extension to current practice is inhibited by caution about the generally low level of yield in the Friesians (approximately 4200 kg milk in 290 days, parity 2) and the use of a complete pelleted diet ('AA6') which is likely to have had characteristics which are far different from the silage based, or even hay based, systems which are the norm in practice. The Gibson (1986) data take no account of differences in feed system.

Since the mid-1970s a selection programme has been underway at the Edinburgh School of Agriculture (Langhill Dairy Research Unit) to investigate the consequences of selection for fat plus protein yield on various aspects of the cow's biological and economic performance. Cows are bred either to top AI proven bulls (on the basis of Improved Contemporary Comparison (ICC) for fat plus protein yield) - the selected line - or to AI bulls of average ICC for fat plus protein - the control line. Initially the control bulls were a panel of 50 young bulls undergoing progeny evaluation in the UK in 1975; from 1986 the control panel has been of average ICC83 but predominantly Holstein.

Until Autumn 1988 only one feeding system was used which was based on a complete mixed diet of high nutrient density, using grass silage, wet brewers grains and concentrates. Concentrate consumption per mature cow was of the order of 2.4 tonnes concentrate per year.

Food intake recording, via Calan-Broadbent gates, was started in 1980 with 46 cows per year; this was increased to 86 cows per year from 1984 and, since 1990, 120 full lactation animal intake records are being taken each year. There is therefore, now, a unique and accumulating database which can be used to explore various aspects of dairy cow performance against a background of well-defined genotype.

Since 1988/89 we have begun to investigate the impact of feed system on the consequences of genetic selection for performance. This has meant adjustment to the feeding regime such that half of each of the control and selected lines of cows (for which food intake is measured) now receive a high concentrate feeding regime, and the other half a high forage feeding regime. On a dry matter basis the grass silage: wet brewers grains: concentrate proportions in the complete mixed diets, averaged over the full lactation, are:

High concentrate - 50:5:45
High forage - 75:5:20

Our aim in the present study is to evaluate the performance, in biological and economic terms, of the two genetic lines in these two feed systems over the first 3 consecutive lactations and clearly, at this stage, the data for that purpose are incomplete. However, information from the accumulated data for the high input feeding system (from 1980), together with results of the first 2 years of the genotype x feed system study (1989-present) have yielded important results about the reliability of genetic indexing and the relationships between genetic index, food intake, feed efficiency and profitability. These can be summarised as follows:

Confirmation of the reliability of sire ICCs for predicting daughters' production. Actual increases in yield of milk and of fat plus protein (kg) have been 99 per cent and 90 per cent respectively of that which would be expected from the ICCs of sires used in the herd (Persaud et al, 1990).

Rates of achievable genetic progress. Over a period of time, the rate of increase in performance of a herd is a consequence of improvements due to breeding, feeding and other aspects of management. Within the Langhill herd we can separate improvements due to breeding from other aspects of management by comparing progress in the control herd with that in the selection herd. The overall rate of improvement in performance has been about 4% per year in the selected herd. The part of this which is due to selection is complex to estimate but approaches half of this total rate.

Associations between Cow Genetic Index, feed efficiency and profitability. Cattle of high genetic index have, on average, a higher feed efficiency (measured as milk energy (MJ) per MJ metabolisable energy (ME) consumed) and are more profitable than their lower index counterparts. Cow Genetic Indexes (CGIs - CGI83 for the current base) are a measurement of the genetic merit of cows for fat and protein production and have been available for predigree recorded animals since the early 1980s. Current average CGI for the UK is around 450 which is also reasonably close to the average CGI of cows in the control line at Langhill. The Langhill selection herd, which has the highest average CGI of any herd in the UK, currently averages around 750. Table 4.5 shows the expected physical and financial performance of heifers and cows (parity 3-5) of different predicted CGI based on Langhill results from the high input feed system used since 1980.

Table 4.5. EXPECTED PHYSICAL AND FINANCIAL PERFORMANCE OF HEIFERS AND COWS OF DIFFERENT PREDICTED CGI OVER 38 WEEKS OF LACTATION (BASED ON LANGHILL RESULTS)

Average	*Heifers*			*Cows*		
Predicted CGI	*450*	*650*	*(% Diff)*	*450*	*650*	*(% Diff)*
Milk (kg)	5225	5849	(11.9)	6997	7621	(8.9)
Fat + Protein (kg)	423	460	(8.7)	546	583	(6.8)
Dry matter intake (kg)	4082	4175	(2.3)	4741	4834	(2.0)
Efficiency (Milk MJ/MJ ME)	0.352	0.375	(6.5)	0.392	0.415	(5.9)
Margin over all food costs (£)	545	648	(18.9)	813	915	(12.5)

This shows the expected (predictable) difference in output of milk and milk fat plus protein and that the higher index, high yielding animals eat more food but are more feed efficient (MJ/MJ) than lower CGI animals. In economic terms high CGI animals outperform their lower CGI counterparts substantially. The advantage is currently worth around 50 pence per predicted CGI point per lactation. [Here it is worth noting that 2 kinds of CGI can be referred to. **Predicted** CGI is based only on information on the heifer's sire and dam. **Actual** CGI also includes the individual's own performance up to lactation 5. The difference in margin over all food costs (Table 4.5) between high and low index animals is about twice as great (ie. about £1 per point) when based on an **actual** CGI rather than that which is predicted].

Options for selection for efficiency. The genetic correlation between fat plus protein yield and feed efficiency is less when all feed is available *ad libitum* than has previously and frequently been assumed. Freeman (1975) used his estimates for heritability of feed efficiency (milk energy per unit feed energy) from a variety of sources and combined that information with estimates of the genetic correlation between feed efficiency and milk yield to conclude that selection for milk production yielded a correlated response in efficiency which was 70-95 per cent of that which would be achieved through direct selection on efficiency. But the sources of information from which this conclusion was derived all used a feeding regime which included an element of feeding to yield. This obviously would impose a closer association between food intake and milk yield than would be the case if food was simply available *ad libitum.*

In our studies at Langhill complete mixed diets have been made available *ad libitum* with no element of feeding to yield. Under those circumstances we find that correlated responses in feed efficiency, following selection for fat plus protein yield are likely to be only 47-74 per cent as great as those for direct selection on efficiency. Thus, at least under *ad libitum* feeding there appears to be a substantial advantage in having estimates of food intake if the selection objective is to improve feed efficiency.

Performance of different genotypes in high and low input feed systems (Langhill data)

Results from the first 2 years of the genetic line/feed system study at Langhill are presented in Table 4.6. Performance data have been regressed on CGIs to allow presentation of the data in the table for animals of average CGI 500 (close to national average and Langhill control herd) and 700 (top 5 per cent of UK cows; similar to Langhill selection herd average).

Table 4.6. PERFORMANCE OF DAIRY COWS OF AVERAGE (CGI = 500) OR HIGH (CGI = 700) GENETIC INDEX SUPPORTED BY A HIGH CONCENTRATE (AVERAGE 450g CONC. DM/kgDM) OR HIGH FORAGE (AVERAGE 200g CONC. DM/kgDM) FEED SYSTEM OVER 38 WEEKS OF LACTATION. DATA ARE ESTIMATED FROM TWO YEARS STUDY AT LANGHILL (1988/9AND 1989/90)

Predicted CGI	*High Concentrate*		*High Forage*	
	700	*500*	*700*	*500*
Milk yield (kg)	7032	6164	6030	5309
Fat (g/kg)	42.5	44.8	46.2	45.1
Protein (g/kg)	30.4	31.5	30.4	31.3
Fat + Protein (kg)	512	466	463	406
Dry matter intake (kg)	4645	4408	4065	4010
Efficiency (Milk MJ/MJ ME)	0.388	0.366	0.414	0.367
Margin over all food costs (£)	759	643	732	599

As indicated previously, high genetic index animals outperform their lower index contemporaries in the high concentrate feeding system. But here we see that this is the case for both systems of feeding as regards output of milk, energetic efficiency and margin over feed costs. Taken over 38 weeks of lactation dry matter intake in high index animals in the high input feed system exceeded that of low index animals by some 5 per cent. But in the high forage feed system, dry matter intake changed relatively little. With the somewhat limited data available to date, this interaction between genotype and food intake is not statistically significant and the situation will become clearer as data accumulate over the planned duration of the study. The constraints to food intake which are imposed by the high forage feed system are, however, obviously material to the manner in which animals of different genotype may respond to lower input systems. From condition score assessment one sees, as expected, that at the end of week 38 of lactation, high CGI animals were, on average, in lower condition than their low CGI contemporaries; and across genotypes, average condition was lower with the high forage than the high input system of feeding - although in each instance the difference was only about 0.2 condition score units (on a 5 point scale)

Of particular interest from Table 4.6 is the comparative performance of high CGI animals in a low input system compared with average CGI animals in the higher input system. For these two categories, output of milk solids (fat plus protein) was essentially identical but food intakes (especially of concentrates) were quite different as were gross energetic efficiencies and margin over food costs which differed by approximately £89 in favour of the high index group supported by the high forage regime.

Consequences for nutrition of future genetic progress

As genetic progress continues the trends indicated in Table 4.6 can also be expected to continue. At present, the highest predicted CGI of animals at Langhill is just in excess of 1100 and increasing numbers of animals of this very high level of index can be expected to enter the UK National Herd in the future; our animals will enter the milking herd in 1993. It is interesting to speculate on the implications from the regressions implied in Table 4.6 for the nutritional support of animals at the high genetic index levels which can be expected in the future.

We have used the regressions underlying Table 4.6 to predict the intake and lactational performance of cows with CGI 1100 in each of the feed systems being evaluated at Langhill. From this, we can estimate a predicted energy deficit for animals under these circumstances. By interpolation between the two feed systems (and a little extrapolation beyond the high input system) we have estimated the likely levels of food intake, lactational performance and predicted energy deficit for a range of feed systems which are characterised according to the metabolisable energy concentration (M/D; MJ ME/kg DM) of the total diet.

Having estimated a projected energy balance for animals of different genotype in different systems of feeding (characterised by M/D), we can then use this to identify dietary energy densities which would need to be provided (for animals of different index) to achieve predicted energy balances which are 'acceptable'. Table 4.7 presents a summary of these calculations using two criteria of assessment for 'acceptable condition'; the first is the condition (ie. achieved calculated energy balance) observed with high CGI animals in the Langhill high concentrate input system of feeding. The second is the condition (predicted energy balance) achieved by selected animals in the high forage input system of feeding. The data are presented as limits to ME intake which might be expected for animals of CGI 500, 700, 900 and 1100. The 'boundaries of acceptability' for intakes to allow achievement of calculated energy deficit as identified above are indicated.

From this, one can see that, as CGI continues to increase, minimal standards of nutritional support (represented as energy concentration of food) will also need to be increased if acceptable degrees of energy deficit are to be accommodated.

Table 4.7. PROJECTED INTAKES OF METABOLIZABLE ENERGY (GJ ME PER 38 WEEKS OF LACTATION) WHICH COWS OF DIFFERENT CGI AND OFFERED DIETS OF VARYING M/D(AVERAGED OVER THE 38 WEEK PERIOD) MAY BE CAPABLE OF ACHIEVING. THE SOLID (—) AND DASHED (- - -) LINES SHOW THE MINIMUM M/D WHICH WOULD NEED TO BE OFFERED TO ALLOW COWS TO ACHIEVE THE (CALCULATED) ENERGY DEFICIT ACTUALLY ACHIEVED BY COWS OF CGI 700 OFFERED THE HIGH CONCENTRATE (—) OR HIGH FORAGE (- -) FEED SYSTEM AT LANGHILL

	CGI			
M/D	500	700	900	1100
11	44.1	44.7	45.3	45.9
11.2	46.0	47.2	48.3	49.4
11.4	48.0	49.6	51.3	53.0
11.6	50.0	52.2	54.4	56.6
11.8	52.0	54.8	57.6	60.4
12	54.1	57.5	60.9	64.3
12.2	56.2	60.2	64.2	68.2

Table 4.7 deals with the implications for energy nutrition of continuing progress in genetic improvement as assessed by CGI. One can go further to consider the implications for protein feeding and Table 4.8 shows the concentrations of crude protein in the concentrate proportion of the diet (and the degradability of that protein) which are calculated to be needed to meet the metabolisable protein (MP) needs of animals of different genotype in the two systems of feeding currently used at Langhill. Because of the constraints to intake which apply, as the animal's ability to secrete protein (at an increased F + P yield) increases, so too must the concentration of protein in the concentrate. This is particularly the case with systems of feeding which emphasise use of forage.

To a considerable extent these calculations are little more than statements of the obvious - as genetic merit for milk output increases it would appear that capacity for food intake increases less quickly. While this results in enhanced feed efficiency it also means that the higher quantities of ME and MP which are needed to sustain higher yields have to be squeezed into a relatively smaller amount of dietary DM. Hence we expect energy and nutrient concentrations to increase to meet the demands of higher index animals if body condition is not to be sacrificed - and also if there is no alteration in metabolic efficiency as genetic index rises.

The data in Tables 4.7 and 4.8 should not be taken as absolute - they are merely conjectural, based on extrapolations of existing data. As further data accumulate, we expect to be able to refine these predictions.

Table 4.8. PROJECTED CONCENTRATIONS OF CRUDE PROTEIN (CP; g/kgDM) AND PROTEIN DEGRADABILITY (dg) REQUIRED IN THE CONCENTRATE PORTION OF THE DIET TO MEET THE METABOLIZABLE PROTEIN NEEDS OF COWS OF DIFFERENT CGI SUSTAINED IN THE HIGH CONCENTRATE OR HIGH FORAGE FEED SYSTEM ADOPTED AT LANGHILL (SILAGE CP ASSUMED TO BE 140g/kgDM;dg 0.8)

Feed system	*High Concentrate*			*High Forage*		
CGI	*500*	*700*	*1100*	*500*	*700*	*1100*
Concentrate						
CP(g/kgDM)	175	174	189	196	219	300
dg	0.74	0.74	0.69	0.6	0.59	0.43

Metabolic efficiency

Is there genetic variation in the efficiency of the individual metabolic processes which contribute to the overall lactational performance of dairy cows? The word 'efficiency' is used here to describe either the energy cost of supporting a particular process, or the mass efficiency of converting substrate into product.

An answer to the question is made difficult by the technical problems of measuring metabolic efficiency of discrete processes in the lactating animal. To partition energy costs between maintenance, body tissue and milk secretory processes by calorimetric techniques (Moe, Tyrrell and Flatt, 1971) is difficult enough. To identify the energy costs associated with the synthesis of particular constituents of milk (fat, protein, lactose) has yet to be achieved with any degree of reliability - although various theoretical estimates exist (Table 4.9). Whether there is variation in the efficiency of discrete processes is therefore principally an academic question. The underlying stoichiometry of synthetic processes seems unlikely to be subject to genetic variation - although possibilities exist for there to be variation in the balance of alternative synthetic pathways which may be used (where these exist) or for the relative activities of synthetic and degradative pathways for products which 'turn over'.

Recently, Miller *et al.* (1991) have measured uptake of various nutrients by the mammary gland in cows varying in yield from 5900-13600 kg per 305 day lactation. They concluded that plasma arterial concentrations of milk fat precursors (acetate, non-esterified fatty acids, betahydroxybutyrate and triacylglycerol) are closely associated with the uptake of these substrates by the mammary gland - indicating a strong link between rate of supply and rate of uptake. Uptake of blood glucose (the major precursor of milk lactose) was, however, not determined by plasma arterial concentration but by other factors such as mammary biosynthetic capacity and blood flow. The extent to which these are under genetic or environmental control is itself debatable (see Knight, 1989) but while the data give clues about the driving forces for milk volume separate from concentration of milk fat (amino acid uptake and protein output were not addressed)

Table 4.9. THEORETICAL ENERGETIC EFFICIENCIES (MJ PRODUCT/TOTAL MJ REQUIRED) OF SYNTHESIS OF PARTICULAR MILK CONSTITUENTS

	Fat	*Protein*	*Lactose*
Oldham (1988)	0.83 - 0.89	0.85	0.90 - 0.94
Smith (1988)	0.67 - 0.98	0.76 - 0.87	0.75 - 0.96
Van Es and Van der Honig (1979)	0.7	0.77	0.73

such studies do not give any measure of the extent to which individuals differ in their metabolic efficiency.

Taylor *et al.* (1986a) have suggested that genetic changes in milk yield might be expected to be accompanied by a correlated genetic increase in maintenance requirement because of differences in the efficiency with which metabolisable energy is used for maintenance in different genotypes. Milligan and Summers (1986) have argued that ion transport (represented by the activity of Na+ K+ ATPase (the sodium pump)) represent a major contribution to maintenance energy expenditure and that this might be genetically determined. We have failed to find any close association between sodium potassium ATPase activity in erythrocyte membranes and CGI in dairy cattle (Pine *et al.*, 1992) although there are indications that the activity of the pump might reflect a degree of balance or imbalance between genetic merit and nutrient provision in the animal (Scollan *et al.*, 1992) such that pump activity is enhanced when genetically determined demand and nutrient provision are not in line. Under such circumstances maintenance efficiency may be reduced in higher merit genotypes if food quality is moderate - which would be an interesting possibility to explain the Taylor *et al.* (1986a) comments.

Grainger *et al.* (1985b) used energy and nitrogen balance experiments with high and low breeding index New Zealand Friesian cows. They found differences in partition of energy according to genotype but, from their data, there appears to be no reason to think that genotype was influencing any partial efficiencies. Using a statistical approach developed by Walter and Mao (1989), Buttazzoni and Mao (1989) estimated variability and heritability of net energy efficiency traits (for maintenance, milk and body weight change). By this approach they identified variation as being least for maintenance, substantially greater for lactation and most variable for body weight change. This would be expected on the grounds that the composition of weight change is notoriously difficult to estimate and, given the possible variation in energetic efficiency according to milk composition (Table 4.9), one might expect variability which would be associated with variation in the compositions of the milks produced by the cows in the study. That variance in net energy efficiency for maintenance was less than one tenth of that for lactation gives perspective to the issue of possible contribution of variation in maintenance efficiency.

A presumptive conclusion from the Buttazzoni and Mao (1989) approach is that there is no real genetic variation in the stoichimetric efficiencies of synthesis/degradation of milk or body constituents - but that there may be variation in the balance of products which contribute to milk or body weight change, such that the net energy efficiency associated with these processes appears to vary. That there is, apparently, variation in maintenance efficiency is interesting - but the variability appears to be relatively small.

Thus the main influence of genotype on overall efficiency is the result of differences between genotypes in the partition of nutrients and energy. As already observed, there are both between- and within-breed differences in nutrient and energy partition which lead to differences in overall feed efficiency - at least some of these implying GxN interactions (Oldenbroek, 1988). The origins of the driving forces which result in differences between individuals either in short term (partial lactation) or long term (full or multi-lactation) efficiency of feed use are the subject of continuing interest. Interest in the role of the somatotrophic axis has been particularly great (Hart, 1988; McBride *et al.*, 1989).

Simple associations between circulating concentrations of blood hormones and genetic index have not been very revealing but both male and female cattle of high predicted breeding value for milk yield are more responsive to standardised challenges with either growth hormone releasing factor (GRF) or thyrotrophin releasing hormone (TRF) than their lower predicted breeding value contemporaries (Lovendahl *et al.*, 1991). This means that, in response to a standard challenge, high breeding value animals appear to secrete more GH than their lower merit contemporaries. Whether target tissues also vary in sensitivity to enhanced hormone secretion in different genotypes is not clear.

Genetic differences in responses to nutrients

Established wisdom (ARC, 1980) has it that responses to a change in feed allowance depend on current yield. It has been argued that the pattern of response to energy supplementation varies according to whether food is available *ad libitum* or not (Broster and Thomas, 1981) - responses to incremental changes in energy intake being similar between 'high' and 'low' yielding phenotypes when food is available *ad libitum* because higher yielders are expected to eat more food. Gordon's (1984) results support this by showing that the difference in yield between a 'high yielding' and a 'low yielding' group of animals in terms of milk yield was essentially the same over a range of concentrate consumptions from 0.6 - 1.6 tonnes for a winter feed period. But whilst the higher yielding animals did eat a little more silage, the difference in silage intake would account for only about two thirds of the difference in milk yield - and, not surprisingly, at the end of the winter feeding period, the high yielding group were in lower average condition score than the low yielders (and this difference was most exaggerated with the lowest concentrate input).

To date there are no controlled experiments which allow interpretation of the responsiveness of different genotypes to controlled additions of nutrients - or, indeed, to the balance of nutrient provision (eg. metabolisable protein:metabolisable energy; form of dietary carbohydrate etc).

The kinds of relationship which led to Table 4.7 above show the possibility of relating food intake capacity, nutrient and energy requirements to genetic index. They might be developed to allow predictions of response to be related to genetic index, which may be helpful to accommodate some of the difficulties in predicting partition of nutrients which exist (Oldham and Emmans, 1988).

Prospects for gene insertion and GxN interactions

A brief comment on the future possibilities for genetic manipulation through gene insertion is justified. The technical (Clark, 1990) and ethical issues associated with gene transfer are considerable. Manipulation of milk (protein) composition does appear to be technically achievable (Simons *et al.*, 1990; Jimenez-Flores and Richardson, 1990). If the concentrations and kinds of protein secreted in milk were changed substantially then one might expect, through conventional calculation, that requirements for supply of amino acids from nutrients in feeds may also change.

An issue of interest to the nutritionist would be the prospect of inserting gene constructs into animals in which the promoter was responsive to a particular nutrient. In theory at least it would be possible to control expression of a gene (for example, coding for GH) by either including, or not, a particular 'nutrient' which would promote expression. Such a 'nutrient' would, of course, have to be an abnormal dietary constituent. If such technology were, however, to progress to the point of application, then genotype x 'nutrient' interaction would be substantial.

Conclusion

Over the last 3 decades advances in nutritional understanding and application through feeding management have yielded considerable improvements in lactational performance in dairy cows. Whilst genetic improvement, through the use of AI, has been an available technology throughout that period, it is the case, in the UK, that only in the last 5-10 years have improvements in performance through breeding begun to be seen at a substantial level. In prospect, however, we have an acceleration in genetic improvement, both through adoption of established technologies and through the introduction into practice of new ones.

Improvements in future performance are therefore likely to be determined at least as much by genetic improvement as by any other management practice. Current evidence suggests that genotype x nutrition interactions are relatively small - but they do exist. Of particular importance is the variation amongst genotypes in food intake capacity and especially capacity to ingest forages. Views differ on the importance of assessing food intake as a trait for future selection (contrast Moore *et al.*, 1990 with Persaud *et al.*, 1991). But an initial analysis of data emerging from the Langhill genotype x feed system study would suggest that we are now at a point in genetic improvement at which interactions between genotype and the capacity of high index animals to accommodate the nutritional limitations of both reduced and higher input feed systems and to remain fit, is open to challenge. From our studies selection for performance using current Cow Genetic Indexes results in more biologically and economically efficient performance. Whilst this applies equally to a lower as well as a higher input feed system, extrapolation to yet lower input feed systems would need to be done with caution.

Nutritional support of high genetic index animals requires thought and consideration. Nutrient:energy ratio in particular needs careful attention.

In answer to the three questions posed at the end of the introduction to this paper we can say:-

1. that selection criteria for animals in lower input systems should recognise the capacity of dairy cattle for food, especially forage, intake

2. multi-trait selection indexes should recognise food intake and feed efficiency as traits and

3. nutritional support of future improved genotypes will need to be carefully targeted.

In cows, genotype x nutrition interactions exist and may become more important. In practice, interactions between geneticists and nutritionists are certainly important - and in many respects long overdue.

Acknowledgements

Thanks are due to SOAFD, AFRC and the Holstein-Friesian Society of Great Britain and Ireland for supporting research at Langhill. Mr R Veerkamp gave valuable assistance in extracting specific data for this paper.

References

ARC (1980). *The Nutrient Requirements of Ruminant Livestock.* CAB Farnham Royal.

Brigstocke, T.D.A., Lindeman, M.A., Cuthbert, N.H., Wilson, P.N. and Cole, J.P.L. (1982). A Note on the Dry Matter Intake of Jersey Cows. *Animal Production*, **35**, 285–287

Broster, W.H. and Thomas, C. (1981). The Influence of Level and Pattern of Concentrate Input on Milk Output. pp. 49–69. In *Recent Advances in Animal Nutrition.* (Ed. W. Haresign). Butterworths, London

Bryant, A.M. (1986). Effect of Genotype on the Performance of Jersey Cows and Per Hectare Production of Milk Solids. In *British Cattle Breeders Club Digest. No. 41*, 55–62

Burnside, E.B. (1991). In *British Cattle Breeders Club Digest. No. 46*

Buttazzoni, L. and Mayo, I.L. (1989). Genetic Parameters of Estimated Net Energy Efficiencies for Milk Production, Maintenance and Body Weight Change in Dairy Cows. *Journal of Dairy Science*, **72**, 671–677

Clark, A.J. (1990). Transgenic Animals Applied to Livestock Production. pp. 37–40. In *Proc. 4th World Congress on Genetics Applied to Livestock Production.* Vol. 14 (Eds. W.G. Hill, R Thompson and G A Woolliams)

Coffey, M. (1991). Personal Communication and Holstein Friesian Society Journal, 1992

Freeman, A.E. (1967). Genetic Aspects of the Efficiency of Nutrient Utilisation for Milk Production. *Journal of Animal Science*, **26**, 976–

Freeman, A.E. (1975). Genetic Variation in Nutrition of Dairy Cattle. In *The Effect of Genetic Variance on Nutritional Requirements of Animals*, pp. 19–46. National Academy of Sciences, Washington, D.C.

Gibson, J.P. (1986). Efficiency and Performance of Genetically High and Low British Friesian and Jersey Cattle. *Anim. Prod*, **42**, 161–182

Gordon, F.J. (1984). The Effect of Level of Concentrate Supplementation Given With Grass Silage During the Winter on the Total Lactation Performance of Autumn-calving Dairy Cows. *J. Agric. Sci. Camb.*, **102**, 163–179

Grainger, C., Davey, A.W.F. and Holmes, C.W. (1985). Performance of Friesian Cows With High and Low Breeding Indexes. 1. Stall Feeding and Grazing Experiments and Performance During the Whole Lactation. *Anim. Prod*, **40**, 379–388

Grainger, C., Holmes, C.W. and Moore, Y.F. (1985). Performance of Friesian Cows With High and Low Breeding Indexes. 2. Energy and Nitrogen Balance Experiments with Lactating and Pregnant, Non-lactating Cows. *Anim. Prod.*, **40**, 389–400

Hart, I.C. (1988). Altering the Efficiency of Milk Production of Dairy Cows with Somatotrophin. pp. 232–247. In *Nutrition and Lactation in the Dairy Cow.* Ed. P.C. Garnsworthy. Butterworths, London

Illius, A.W. and Gordon, I. (1991). Prediction of Intake and Digestion in Ruminants by a Model of Rumen Kinetics Integrating Animal Size and Plant Characteristics. *J. Agric. Sci., Cambs*, **116**, 145–157

Jimenez-Flores, R. and Richardson, T.l (1990). Future Possibilities Using Biotechnological Methods to Improve Milk Quality. pp. 726–737. In *Proceedings of the 23rd International Dairy Congress. Vol. I.* International Dairy Federation, Brussels

Knight, C.H. (1989). Constraints on Frequent or Continuous Lactation. *Proc. Nutr. Soc*, **48**, 45–51

Korver, S. (1982). Genetic Aspects of Feed Intake and Feed Efficiency in Dairy Cattle: A Review. *Lives. Prod. Sci*, **20**, 1–13

Lawlor, T.K. and Short, T.H. (1990). The Application of Quantitative Genetic Information Within the United States Holstein Population. pp. 179–182. In *Proceedings of the 4th World Congress on Genetics Applied to Livestock Production.* (Eds. W.G. Hill, R. Thompson and J.A. Woolliams) Vol. XIV. Edinburgh

Lovendahl, P., Angus, K.D. and Woolliams, J.A. (1991). The Effect of Genetic Selection for Milk Yield on the Response to Growth Hormone Secretagogs in Immature Cattle. *J. Endoc*, **128**, 419–424

McBride, B.W., Burton, J.L. and Burton, J.H. (1988). Review: The Influence of Bovine Growth Hormone (Somatotrophin) on Animals and Their Products. *Research and Development in Agriculture*, **5**, 1

Milk Marketing Board (1985). Report of the Breeding and Production Organisation 1984/5 No. 35. Thames Ditton, Surrey

Miller, P.S., Reis, B.L., Calvert, C.C., Depeters, B.J. and Baldwin, R.L. (1991). Patterns of Nutrient Uptake by the Mammary Glands of Lactating Dairy Cows. *J. Dairy Sci*, **74**, 791–799

Milligan, L.P. and Summers, M. (1986). The Biological Basis of Maintenance and its Relevance to Assessing Responses to Nutrients. *Proc. Nutr. Soc*, **45**, 185–193

Moe, P.W., Tyrrell, H.F. and Flatt, W.P. (1971). Energetics of Tissue Mobilisation. *J. Dairy Sci*, **54**, 548–553

Moore, R.K., Kennedy, B.W., Schaeffer, L.R. and Moxley, J.E. (1990). Parameter Estimates for Feed Intake Production in First Lactation Using Milk Recording Data. *J. Dairy Sci*, **73**, 826–834

Oldenbroek, J.K. (1986). The Performance of Jersey Heifers and Heifers of Larger Dairy Breeds on Two Complete Diets With Different Roughage Contents. *Livest. Prod. Sci.*, **14**, 1–14

Oldenbroek, J.K. (1986). The Performance of Jersey Cows and Cows of Larger Dairy Breeds on Two Complete Diets With Different Roughage Contents *Livest. Prod. Sci.*, **18**, 1–17

Oldenbroek, J.K. (1986). Feed Intake and Energy Utilisation in Dairy Cows of Different Breeds. Ph.D. Thesis, Wageningen University

Oldham, J.D. and Emmans, G.C. (1988). Prediction of Responses to Protein and Energy Yielding Nutrients. pp. 76–96. In *Nutrition and Lactation in the Dairy Cow.* Ed. P.C. Garnsworthy. Butterworths, London

Orskov, E.R., Ojwang, I and Reid, G.W. (1988). A Study on Consistency of Differences Between Cows in Rumen Outflow Rates of Fibrous Particles and Other Substrates and Consequences for Digestibility and Intake of Roughages. *Anim. Prod*, **47**, 45–51

Pearson, R., Vinson, W.E. and Mainert, T.R. (1990). The Potential for Increasing Productivity Through Selection for Increased Milk and Component Yields, pp 104–113. In *Proceedings of the 4th World Conference on Genetics Applied to Livestock Production.* Ed. W.G. Hill, R. Thompson and J.A. Woolliams. Vol. XIV. Edinburgh

Persaud, P., Simm, G., Parkinson, H. and Hill, W.G. (1990). Relationships Between Sire's Transmitting Ability for Production and Daughters' Production, Food Intake and Efficiency in a High Yielding Dairy Herd. *Anim. Prod*, **51**, 245–253

Peterson, R. (1988). Comparison of Canadian and New Zealand Sires in New Zealand for Production, Weight and Conformation Traits (Unpublished Report)

Pine, A.P., Jessop, N.S. and Oldham, J.D. (1992). The Influence of Genotype, Diet and Stage of Lactation on Erythrocyte ATPase Activity in Dairy Cattle. *Proc. Nutr. Soc.*, **51**, (in press).

Schneider, H. (1946). Nutrition and Disease. *Vitam. Horm.*, **4**, 60

Scollon, N.D., Zhaw, X., McBride, B.W., Jessop, N.S., Allen, G.F. and Oldham, J.D. (1992). Estimation of NA+, K+, ATPase Activity in Bovine Erythrocytes and the Use of NA+, K+ and ATPase as an Indicator of the Genetic Merit of Dairy Cows. In *Proc. Energy Metabolism Symposium, Zurich.* (in press).

Simm, G., Wray, N. and Persaud, P. (1991). Milk Production, Food Intake and Efficiency - An Update from Langhill. *British Cattle Breeders Club Digest*, No. 46. pp. 40–47.

Simons, J.P., Ali, S., Archibald, A.L., Brown, P., Harris, S., McClenaghan, N., Whitelaw, C.B.A., Wilmutt, I. and Clark, A.J. (1990). Modification of Milk Composition by Gene Transfer. pp. 132-135. In *Proc. 4th World Congress on Genetics Applied to Livestock Production.* Vol. 14. (Eds. W.G. Hill, R. Thompson and J.A. Woolliams)

Smith, N.E. (1988). Alteration of Efficiency of Milk Production in Dairy Cows by Manipulation of the Diet, pp 216–231. In *Nutrition and Lactation in the Dairy Cow.* Ed. P.C. Garnsworthy. Butterworths, London

Sunde, M.L. (1975). Introduction in *The Effect of Genetic Variance on Nutritional Requirements of Animals.* (pp. 1–2) National Academy of Sciences, Washington, D.C.

Taylor, St. C.S., Thiessen, R.B. and Murray, J. (1986a). Inter- breed Relationship of Maintenance Efficiency to Milk Yield in Cattle. *Anim. Prod.*, **43**, 37–62

Taylor, St. C.S., Moore, E.G and Thiessen, R.B. (1986b). Voluntary Food Intake in Relation to Body Weight Among British Breeds of Cattle. *Anim. Prod.*, **42**, 11–18

Van Es and Van der Honig (1979).

Walter, J.P. and Mayo, I.L. (1989). Modelling Net Energy Efficiencies as Quantitative Characteristics for Genetic Analysis in Lactating Cows. *J. Dairy Sci.*, **72**, 2362–2374

Woolliams, J.A. (1990) pp. 15-24. In 'Proceedings of the 4th World Congress on Genetics Applied to Livestock Production' (Eds. W.G.Hill, R.Thompson and J.A.Woolliams) Vol.XIV. Edinburgh

5

THE METABOLIZABLE PROTEIN SYSTEM FOR RUMINANTS

A.J.F.WEBSTER

Department of Animal Husbandry, University of Bristol,
Langford, Bristol, U.K.

Introduction

The objective of any feed evaluation system for ruminants is to predict the supply of available nutrients from robust measurements which can be made in the laboratory yet which recognise the essential principles of ruminal and post-ruminal digestion. At the 1988 Feed Manufacturers Conference four criteria for a successful feeding system were proposed (Webster, Dewhurst and Waters, 1988). These are repeated here in slightly modified form.

1. It should be based on measurements of feed chemistry, physical form or biological degradation that can be adopted as routine by the feed compounder.

2. It should be deterministic and sufficiently descriptive of ruminant physiology to be able to incorporate essentials of present and future knowledge.

3. It should predict the yield of the major truly-absorbed substrates for energy and protein metabolism.

4. It must, when tested in production trials, be demonstrably better than existing empirical systems.

Any feeding system, such as the Metabolizable Energy (ME) System (Agricultural Research Council, ARC 1980) or the Metabolizable Protein (MP) System (Interdepartmental Working Party, IDWP, 1991; Webster, 1987) defines both the supply of a specific nutrient from the gut to the tissues and the requirement for that nutrient to meet the needs of maintenance and a defined level of production (e.g. growth or milk yield). The terms ME and MP define the supply of absorbable nutrients and this is related to requirement using efficiency terms (e.g. k_l and k_{nl} for lactation, k_f and k_{ng} for growth in ME and MP respectively).

The limitations of digestible crude protein (DCP) as a feeding system for ruminants have been recognised for many years. It is acknowledged less widely that these limitations also apply to the ME system as currently applied:

1. Neither DCP nor ME distinguishes between ruminal and post- ruminal digestion.

2. Neither system can define substrate supply, e.g. individual volatile fatty acids or amino acids.

3. Neither system can account for energy/protein interactions either in the rumen or the tissues.

4. Both systems confound biochemical efficiency with production response.

This last criticism requires further explanation. In the ME system k_l and k_f define (in theory) the proportion of energy retained in excess of maintenance but do not directly relate ME intake to milk yield or lean tissue growth. In other words they do not define the marginal response to increasing nutrient supply in terms of yield of saleable product. DCP and its successors (including MP) also attempt to relate protein requirement to supply using terms which define the biochemical efficiency of utilisation of absorbed amino N in circumstances where no amino acid is present in excess of "requirement." This definition is inevitably restricted to circumstances where growth or milk production are themselves restricted, probably to an extent that makes little commercial sense. A practical feeding system should not attempt to impose rigid concepts of nutrient requirement and maximum efficiency of utilisation but explore marginal responses to increments of nutrients (for whatever biological reason) at commercial levels of production within the full range from maintenance to each animal's genetic potential. For protein, it is inherently probable that the optimal production response will occur when the ratio of increasing yield to supply is below the maximum biochemical efficiency of utilisation of absorbed protein.

Development of new protein systems

All new protein systems distinguish (1) organic N that is degraded in the rumen (RDN) and which can be incorporated into microbial N to an extent that is determined by the supply of fermentable energy; (2) protein N which escapes degradation in the rumen (UDN). Alderman (1987) has collated all the factors adopted by different groups in the construction of feeding systems from these two simple, agreed premises. Table 5.1 examines the most important similarities and differences among the new systems in use in France (Proteins digestibles dans l'intestin, P.D.I., Institut Nationale Recherche Agronomique, INRA, 1988, Verite *et al.*, 1987), Scandinavia (Madsen, 1985) and the USA (National Research Council, NRC 1985) and those proposed for UK by ARC (1984) and IDWP (1991). All systems except ARC (1984) assume that the incorporation of RDN into microbial crude protein (MCP) cannot achieve an efficiency of 1.0 unless RDN is deficient - in which case measurement of efficiency is confounded even more than usual by recycling of urea from tissue metabolism in the host animal. The decision of IDWP to partition RDN into quickly and slowly degraded fractions (QDN and SDN) will be discussed later when the MP system is examined in more detail.

Table 5.1. COMPARISON OF CONCEPTS AND FACTORS IN NEW PROTEIN SYSTEMS FOR RUMINANTS

	Protein system				
				UK	
	France	*Scand.*	*USA*	*ARC*	*IDWP*
"Effective RDN	0.9RDN	var.	0.9RDN	1.0RDN	0.8QDN 1.0SDN
Fermentable energy[1]	FOM	DCHO	TDN	ME	FME
True digestibility of UDN	var.	0.82	0.82	0.85	var.
Efficiency of utilisation of TAAN					
lactation	0.64	0.73	0.65	0.80	0.68
growth	0.60	-	0.50	0.80	0.59
CP requirement[2], 30kg milk/day					
g/kgDM	162	157	168	129	173
optimal degradability	0.60	0.68	0.60	0.75	0.66

For full abreviations, see text
[1] FOM = DOM - fat - UDP - products of fermentation
DCHO = DNFE + DCF
FME = ME - ME_{ferm} - ME_{fat}
[2] Values for a 600kg cow yielding 30kg milk and losing 0.5kg bodyweight per day.

The energy required for synthesis of microbial crude protein (MCP) entering the abomasum can, of course, only come from organic matter fermented in the rumen. The USA and ARC systems assume this to be a fixed proportion of total digestible nutrients (TDN) or ME respectively. The Nordic system is based on digestible carbohydrate, (dCHO), the French on fermentable organic matter (FOM) and MP on fermentable ME (FME). Each of these distinguishes between fermentable and unfermentable sources of energy, the revised French system (Verite 1987, INRA, 1988) being the most comprehensive, namely FOM = DOM-fat-UDP - products of fermentation, where DOM = digestible organic matter, UDP = undegraded dietary protein. The Nordic system is based on digestible crude fibre (DCF) and N-free extract (DNFE) which ignores the relatively small contribution of protein to ATP synthesis by microbes (Tamminga, 1982). IDWP ignore UDP. The "newer" systems, IDWP and the revised PDI system, recognise that the digestibility of UDN cannot be taken as a constant.

All systems calculate the yield of truly absorbed amino N (TAAN) from microbial crude protein and UDN then apply efficiency terms to relate TAAN to protein retention during growth of body tissue (or wool) or to protein synthesis in milk. France and the U.K. have achieved concorde here, assuming 0.64 and 0.68 respectively for k_{nl} and 0.60 and 0.59 for k_{ng}, although, as will be argued later, these values may not be very useful in practice.

Table 5.1 also calculates the combination of optimal degradability and dietary crude protein concentration required to meet the protein requirement of a 600kg cow yielding 30kg milk and losing 0.5kg body weight per day. Values are from Alderman (1987) and IDWP (1991). It is now generally accepted that the ARC (1980) estimates are unrealistically low. The concordance between other systems is quite close. IDWP calls for the greatest concentration of crude protein (173 g/kg DM) but this relates to a ration (Table 5.2) which includes 0.55 DM as silage with a relatively high proportion of QDN.

The Metabolizable Protein System

The MP system, as proposed by IDWP (1991), will now be described in more detail, indicating why it is the best that can be done in the light of current knowledge; its use in ration formulation will be illustrated then its limitations will be critically examined and suggestions made as to how they may be overcome.

The first term that requires explanation is Metabolizable Protein itself. IDWP have accepted the principle of ARC (1980, 1984) for a system based on TAAN. It recognises however that the ruminant feed industry has lived comfortably and successfully with the concept of crude protein (N x 6.25) and sees no reason to break this association simply to achieve an (incomplete) sense of semantic respectability. Metabolizable Protein is therefore 6.25 TAAN. The title also preserves a consistency with Metabolizable Energy, partly out of respect to the late Sir Kenneth Blaxter, but mainly because the logic of MP, as illustrated by Figure 5.1, is essentially similar to that of the ME system. All the equations and assumptions that make up the MP system are contained in IDWP (1991). Reference will only be made here to those which are novel in concept or which require further discussion. Effective rumen degradable protein (ERDP), expressed as g/kg DM, is calculated from crude protein. QDP corresponds to a, or degradability at zero time in the equation of Orskov and McDonald (1979) and is assumed to convert to MCP at an efficiency of 0.8, that proposed by ARC (1980) for non-protein N. The degradability of SDP is defined by rumen outflow rate (r, h-1), where r = 0.02, 0.05 and 0.08 for maintenance, growth and lactation respectively. The efficiency of conversion of SDP to MCP is taken as 1.0, thus

$$[\text{ERDP}] = [\text{CP}] \times (0.8a + bc/(c + r)) \tag{5.1}$$

where a, b and c are the fitted parameters for the Orskov and McDonald (1979) equation.

Digestible undegraded protein (DUP,g/kg) is given by

$$[\text{DUP}] = ([\text{CP}] \times (1 - a - bc/(c + r))) - 6.25[\text{ADIN}] \tag{5.2}$$

This equation assumes that acid detergent insoluble N (ADIN) is completely indigestible and that the true digestibility of the remaining Nitrogen is 0.9 (Webster *et al.*, 1988). Fermentable ME (FME, MJ/kg DM) is given by

$$[\text{FME}] = [\text{ME}] - [ME_{fat}] - [\text{ME}_{\text{ferm}}] \tag{5.3}$$

Here ME_{fat} can be taken as 35 MJ/kg fat and ME_{ferm} as 0.1[ME] for ensiled feeds. This latter approximation is probably acceptable for ensiled forage crops but may overestimate

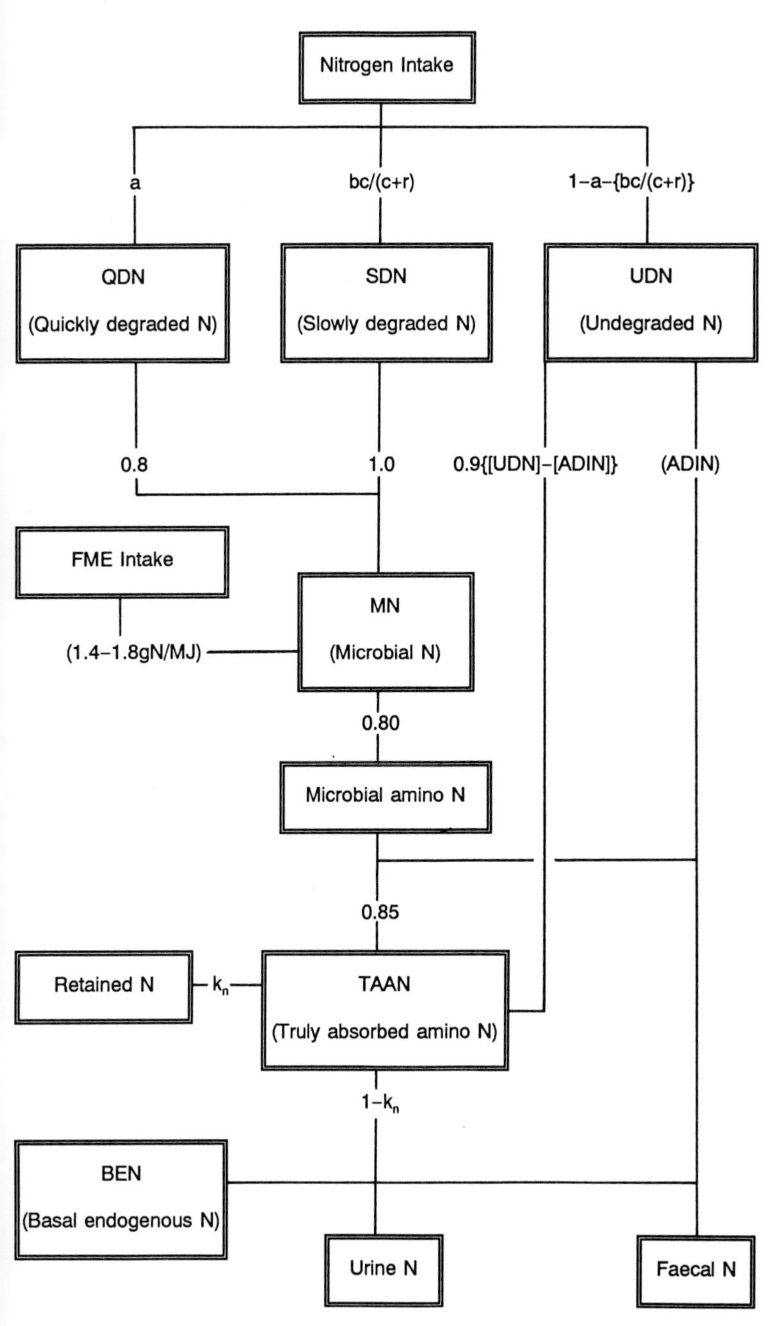

Figure 5.1. Proposed scheme for the description of protein value of diets in terms of truly absorbed amino nitrogen

the concentration of fermentation products in ensiled brewers or distillers grains for which 0.05[ME] may be more appropriate. (C.J.Hyslop, personal communication).

The factors relating microbial protein yield (MCP,g/d) to FME, Ymcp/fme are as follows

Ymcp/fme = 9g MCP/MJ FME at maintenance
10g MCP/MJ FME for growth
11g MCP/MJ FME for lactation

These values correspond to rumen fractional outflow rates of 0.02, 0.05 and 0.08 h-1 respectively.

Metabolizable protein supply (MP g/d) then becomes

$$\mathrm{MP} = 0.68\mathrm{MCP} + \mathrm{DUP} \tag{5.4}$$

The factor 0.68 is that of ARC (1984), based on the assumption that 0.80 MCP is present as true protein and the true absorbability of amino acids is 0.85. Requirement for MP (MPR,g/d) is defined by net protein requirement for maintenance, equivalent to 6.25 basal endogenous N loss (BEN) and net protein gain in milk, liveweight, wool, etc. each divided by its net efficiency term. Thus for a dairy cow neither gaining nor losing weight

$$\mathrm{MPR} = \mathrm{NP_b}/\mathrm{k_n b} + \mathrm{NP_d}/\mathrm{k_{nd}} + \mathrm{NP_l}/\mathrm{k_{nl}} \tag{5.5}$$

where NP_b, NP_d and NP_l are net protein requirements for maintenance, synthesis of scurf and hair and milk production respectively and k_{nb}, k_{nd} and k_{nl} are the corresponding factors for net efficiency of utilisation of MP.

Basal endogenous N (BEN) is taken as $0.35W^{0.75}$ gN/d. This corresponds to the ARC (1984) value for total endogenous N loss in cattle and sheep with no microbial activity in the rumen and sustained entirely by infusion of nutrients, i.e. circumstances where recycling of endogenous urea to amino acids via the rumen microbes is impossible. This value is considerably higher than actual basal losses of endogenous N via faeces and urine (Owens, 1982).

The MP system does not formally incorporate a pathway for recycling urea to the rumen although it recognises recycling in the assumptions used to calculate ERDP (equation 5.1). It also has no difficulty in accommodating the fact that non-ammonia N (NAN) flow to the duodenum may frequently exceed intake of dietary N, due to incorporation of recycled endogenous N. The efficiency terms (k_{nl}, k_{ng} etc.) and BEN are both based on the assumption that recycling does not occur. Any system that incorporates measurements of NAN flow at the duodenum or true excretion of endogenous N must also incorporate an accurate estimate of endogenous N and reduce the efficiency terms k_{nl}, k_{ng} etc. accordingly. The MP system circumvents these hazards of double accounting by adopting the simple maxim that the only true input of crude protein to a ruminant is through the mouth.

Table 5.2. DIET FORMULATION FOR A DAIRY COW YIELDING 30kg MILK AND LOSING 0.5kg LIVEWEIGHT PER DAY (FROM IDWP, 1991)

	Composition/kgDM				
Feeds available	*ME*(MJ)	*FME*(MJ)	*CP*(g)	*ERDP*(g)	*DUP*(g)
Grass silage	10.3	7.8	174	110	23
Dried beet pulp	12.5	12.3	103	43	43
Rolled barley	12.8	12.3	114	84	18
Maize gluten	12.7	11.5	207	130	41
Rapeseed meal	12.0	10.8	400	265	78

MP requirement/day		
	600kg cow, maintenance =	328
	30kg milk	1341
	less 0.5 liveweight loss	-69
		1600

			Intake/day			
Diet	*DM*(kg)	*ME*(MJ)	*FME*(MJ)	*CP*(g)	*ERDP*(g)	*DUP*(g)
36kg grass silage	9.0	93	70	1556	990	207
3.7kg dried bet pulp	3.2	40	39	330	138	138
2.3kg rolled barley	2.1	27	26	239	175	38
2.3kg maize gluten	2.0	25	23	414	260	82
1.4kg rapeseed meal	1.2	14	13	480	318	94
Totals	17.5	199	171	3029	1882	559

Then ERDP/FME = 1882/171 = 11.0g/MJ
MP = 0.68(11.0 × 171) + 559 = 1838
MP/MPR = 1.14 CP concentration = 173g/kgDM

Diet formulation

IDWP (1991) have illustrated the principle of diet formulation using the example of a 600kg dairy cow, producing 30kg milk of 32g/kg protein and losing 0.5kg liveweight per day. This provides an exact comparison with the other systems illustrated by Alderman (1987) and summarised in Table 5.1. This example is largely self-explanatory but several points are worthy of comment. For grass silage the ratio FME:ME is 0.76 and the proportion of CP potentially available as MP is only 0.56 [(0.68 x 110) + 23]. The MP system gives substantially lower estimates for the protein value of grass silage than either DCP or ARC (1980). MCP is constrained by the low ratios of FME:ME and ERDP:RDP, and the true availability of UDP is only 0.48, due to the high proportion of UDP as ADIN. As we improve both our capacity to make good silage and our understanding of what it is we have done, we are compelled to amend our view of the concentrate portion of the ration. Whereas once it was seen primarily as a source of extra energy far greater importance is now placed on its value as a source of high quality protein. The diet in Table 5.2 is made up of 9.0kg DM as grass silage plus 8.5kg DM of concentrate feed containing dried beet pulp, rolled barley, maize gluten and rapeseed meal. In this mixture, the ratio FME:ME is 0.95, and the proportion of CP potentially available as MP is 0.65.

In the complete diet, the ratio ERDP:FME = 11.0g/MJ which is exactly what the system demands to balance the supply of organic N and fermentable energy to the rumen microbes. The ratio of MP supply to requirement is 1.14 which exceeds the safety margin of 10% recommended by IDWP.

In the overall diet, 0.62 of CP is present as ERDP and 0.18 as DUP. The MP system does not call for large quantities of UDP, indeed it calls for more RDP than other systems, particularly for dairy cows because it assumes a higher capacity for MCP yield (11g/MJ FME). The value of UDP is however strongly influenced by the proportion present as ADIN.

Attractions of metabolizable protein

The chief attractions of MP, in common with the revised French and Nordic systems, are that:

1. It relates microbial yield to fermentable energy.

2. It permits flexibility in defining ERDP and DUP.

3. It appears to give "sensible" answers for diets fed to dairy cows in the U.K.

Limitations of Metabolizable Protein

In this section the limitations of MP will be discussed. These concern matters of accuracy and uncertainty, some of them very important. However, one of the chief attractions of MP is that it is conceptually sound without being too rigid so that it can accommodate new information as it emerges.

Effective rumen degradable protein

The first limitation of ERDP is that it is still based on measurements of N disappearance from artificial fibre (AF) bags incubated in the rumen. This technique gives alarmingly different answers when tested at different laboratories (Oldham, 1987; IDWP, 1991). It involves the use of fistulated animals and, perhaps most serious of all, it cannot be assumed that N disappearance from the bag is synonymous with protein degradation. Waters (1991) has observed 20% disappearance of ADIN from AF bags in circumstances where ADIN is known to be completely indigestible. Attempts at Bristol to predict ERDP from feed chemistry have had limited success. The equation proved by Webster *et al.*(1988) is reasonably accurate for grass forages.

$$P_n = (0.9 - 2.4r)(CP - 0.059NDF)/CP \qquad (5.6)$$

where P_n is effective degradability of total N and NDF is neutral detergent fibre. However, as yet, no satisfactory equation has been provided to predict P_n for raw materials (other than grasses) or compound feeds (Waters, 1991; Chaudhry, Webster and Marsden, 1992). The second limitation of ERDP derives from the assumptions that QDP equals water-soluble N, and that the efficiency of conversion of water-soluble N to MCP is a constant, 0.8. This element of uncertainty can only be resolved by experiments which (1) define the chemistry of water-soluble N and (2) measure its rate of incorporation into MCP. However, as stated earlier, the MP system can accommodate this new knowledge as it arises.

Yield of microbial crude protein

This is, and will probably remain, the largest element of uncertainty attached to the estimation of protein supply in all the new systems. It is affected by

1. the relative rates of energy fermentation and N degradation in the rumen;
2. the contribution of endogenous N to microbial protein synthesis;
3. the true energy costs of microbial maintenance and protein synthesis in the rumen;
4. the ratio of outflow rate of microbial protein to total protein synthesis.

IDWP (1991) only consider (inadequately) factors 1 and 4 and predict that the yield of MCP will vary within the range 9 to 11g/MJ FME according to circumstances. Corresponding estimates from other national systems, expressed in terms of ME or FME are as follows (IDWP 1991).

French, 8.7 MCP/MJ FME or 7.6 MCP/MJ ME (INRA, 1985)

Nordic, 10.3g MCP/MJ ME (Madsen, 1985)

USA, MCP(g/d) = 10.6 MEI - 193 (NRC, 1985)

Germany, MCP(g/d) = 11.92 MEI - 15.0 DMI (Rohr,1986)

The fixed efficiency assumed by the French system is 0.74 that of the Nordic system. Both the USA and German systems, in common with IDWP, imply an increasing efficiency with increasing plane of nutrition. The NRC (1985) system implies an increase in the ratio MCP/ME from 7.4 to 9.8 with an increase in MEI from one to four times maintenance. The discrepancy between the French and the other three systems can probably be attributed largely to factor 4, i.e. plane of nutrition and so is taken into account by MP. Nevertheless there is considerable scope for variation in factor 3, e.g. in response to ionophores like monensin (Dewhurst and Webster, 1992). The most promising technique for exploration of these sources of variation is that which predicts microbial protein yield from increments of urinary excretion of purines (Chen *et al.*, 1990; Dewhurst and Webster, 1992). In practice, the element of uncertainty in predicting MCP may not be so serious as it appears in theory since DUP in excess of tissue requirement can be recycled as endogenous urea.

Table 5.3. MEASUREMENT OF THE APPARENT DIGESTIBILITY OF N AND ADIN IN FEEDS COMPARED WITH IDWP ESTIMATES FOR THE TRUE DIGESTIBILITY OF UDN

		Apparent Digestibility		*Estimated true digestibility of UDN*[1] *from ADIN in*	
	n	*N*	*ADIN*	*feed*	*faeces*
Conventional diets	25	0.72	0.009	0.62	0.58
Distillers products	7	0.40	0.51	(0.02)	0.49
High tannin feeds	5	0.57	(-0.89)	0.45	(0.02)

[1] dUN = 0.9(UDN - ADIN)/UDN

Digestible undegradable protein

Waters, Kitcherside and Webster (1992) have measured the apparent digestibility of total N and ADIN in 37 diets incorporating a wide range of raw materials used in compound feeds. In these diets the "test" feed constituted practically the sole source of UDN, the other energy constituents being starch and/or nutritionally improved straw and urea as the sole source of additional RDN. These measurements are summarised briefly in Table 5.3 and compared with estimates of the true digestibility of UDN from dUDN = 0.9 (UDN-ADIN)/UDN based on measurements of ADIN in feed and faeces. Distillers products are grouped separately since they are known to contain substantially more ADIN than present in the original grain due to Maillard reactions taking place during prolonged exposure to heat and moisture (Klopfenstein and Britton, 1987; Van Soest and Mason, 1991). The high tannin feeds included salseed, sheanut and maize kernel. The conventional diets represent a wide selection of raw materials used in compound feeds (e.g. cereals, brans, maize gluten, oil seed cakes). Some of these, especially maize gluten feed, may have contained small amounts of "added" ADIN due to Maillard reactions. The first, reassuring conclusion to be drawn from Table 5.3 is that for the majority of "conventional" diets ADIN may be assumed to be undegradable and indigestible. For these diets the estimated true digestibility of UDN (dUDN) was, on average, 0.62 (range 0.48 to 0.83), considerably lower than the value of 0.85 proposed by ARC (1980). In distillers' products the apparent digestibility of total N was 0.40 and that of ADIN 0.51. The IDWP prediction of true digestibility (0.02) based on ADIN in food is clearly wrong, although that of 0.49 based on indigestible ADIN (i.e. ADIN in faeces) is sensible when compared with the value of 0.40 for the apparent digestibility of total N (which of course incorporates MCP and metabolic faecal N). The protein-binding effect of tannins is described, at least in part, by an increase in ADIN en route through the gut, giving meaningless values for digestibility of ADIN. However the estimate of dUN based on ADIN in the original feed is probably valid. It seems therefore that ADIN in raw materials can be used as a basis for prediction of dUN unless that material has been processed in such a way as to increase ADIN concentration substantially above that contained in the original crop.

Table 5.4. DIRECT MEASUREMENTS OF THE TRUE DIGESTIBILITY OF UNDEGRADED DIETARY N FED TO RATS COMPARED WITH ESTIMATE BASED ON ADIN

	Soya bean meal	*Distillers grains*	
		I	*B*
Raw			
Total N(g/kgDM)	86.0	54.6	48.6
ADIN (g/kgDM)	7.3	13.5	20.8
Rumen degradability	0.68	0.82	0.71
Undegraded material in rat diets			
Total N (g/kgDM)	24.2	28.9	28.0
ADIN (g/kgDM)	0.05	10.1	14.0
True digestibility UDN	0.88	0.59	0.22
Estimated dUN in rat diets[1]	0.90	0.58	0.45

[1] from dUN = 0.9(TN - ADIN)/TN in digestion trials with rats fed diets containing UDN as the sole source of N.

The measurements of apparent availability of total N and true digestibility of UDN in Table 5.3 are, of course, not strictly comparable. Abdul Chaudhry, at Bristol, is currently measuring dUN in rats by the direct, if laborious expedient of harvesting UDN from AF bags incubated in the rumen of cows for 18hr, sterilising it by irradiation then incorporating it as the sole source of protein in semisynthetic diets. Table 5.4 presents preliminary results for soyabean meal and two sources of distillers' dark grains (I and B). The true digestibility of soyabean UDN for rats was 0.88 which compares well with the IDWP estimate of 0.90. The true digestibilities of UDN from the two samples of dark distillers' grains I and B were 0.59 and 0.22 respectively, the digestibility decreasing with increasing ADIN (0.25 and 0.43 of total N in diets I and B respectively). Estimated dUN values based on ADIN in the diets as feed to the rats were I = 0.58 and B = 0.45, once again not that divergent from the direct measurements. The effect of "added" ADIN resulting from Maillard reactions on dUN is not yet clear enough to justify any general recommendation. However, the results presented in Table 5.4 do support the conclusion that the nutritive value of added ADIN acquired during processing is still very low and not underestimated by equation 5.2.

Testing the metabolizable protein system

It was stated at the outset that one essential criterion for any new feeding system is that it must, when tested in production trials, be demonstrably better than existing empirical systems. This claim cannot be sustained for MP on the basis of the tests described by IDWP (1991). The first trial involved a co-ordinated dairy cattle experiment at the former National Institute for Research in Dairying (NIRD), Grassland Research Institute (GRI) and Boxworth Experimental Husbandry Farm. The CP concentrations in the concentrates were varied by substituting cereals with fishmeal (high UDP) and soyabean meal (high RDP) at two levels of inclusion. At NIRD and GRI teams of duodenally cannulated cows were offered the same diets as those in the main production trial to relate duodenal supply of amino acids (NAN) to ARC (1980, 1984) predictions and to yields of milk protein. Measured valued of NAN flow at the duodenum did not correspond to those predicted by ARC (1980, 1984) nor show any obvious correlation with yield of milk protein. IDWP concluded "it is a matter for serious and general concern that the basis for proper testing of any 'new' protein system which uses the general concepts of microbial and animal 'protein' needs remains technically imperfect," and rather gave up on this substantial and expensively acquired body of data.

The second test by IDWP of the MP system involved an examination of a large data base for dairy cattle established by Andrew Rook (IGAP, Hurley). This was used to compare MP supply with estimated requirement in the ARC (1980, 1984) and IDWP (1991) systems, after applying an initial screening procedure which ensured that "for at least one of the systems, estimated MP supply was within 10% of established requirements." Because ARC (1980) predicts lower protein requirements than the other two systems, this necessarily implies that for the majority of data identified as "potentially N limiting by one of the systems" N supply will have been less than requirements for both ARC (1984) and IDWP (1991).

It can be argued that this exercise does not constitute a proper test of any of the three systems in practice, partly because the observed differences largely reflect the inherent differences in the internal logic of the three systems but mainly because the data have been preselected to impose conditions where k_{nl} is maximal, i.e. where MP is limiting and this may not correspond to responses to MP when high-yielding cows are fed high concentrations of CP in practice.

Van Straalen (1991) has compared yields of milk protein (Y_p) in practical feeding trials to estimates of metabolizable protein supply as defined by DCP, NRC (1985), INRA

Table 5.5. RELATIONSHIP BETWEEN OBSERVED YIELD OF MILK PROTEIN (Y_p, g/d) AND METABOLIZABLE PROTEIN SUPPLY (kg/d) AS DEFINED BY DIFFERENT NATIONAL SYSTEMS

System	*regression*	R^2	*MSEP*[1] (%)
Net energy lactation	$Y_p = 230 + 36NER(MJ)$	0.72	-
DCP	$Y_p = 675 + 104DCP$	0.12	4.3
USA(NRC 1985)	$Y_p = 431 + 210MP$	0.43	9.4
France (PNRA 1988)	$Y_p = 428 + 241MP$	0.55	4.8
Nordic (Madsen 1985)	$Y_p = 367 + 273MP$	0.55	10.6
UK (ARC 1984)	$Y_p = 389 + 491MP$	0.43	33.8

[1]MSEP = mean square error of prediction

(1988), Madsen (1985) and ARC (1984). His observations are summarised in Table 5.5. The first observation is that Y_p correlated better with net energy intake (NEl) than with intake of MP as defined by any system. The correlation of Y_p with DCP intake was very poor (which justifies the attempts to replace it). All systems overestimated the actual efficiency of conversion of MP to Y_p, the greatest error being that of ARC (1984) for which the mean square error of prediction was 33.8%. One interpretation of Van Straalen's (1991) data is that the large positive intercepts and range of coefficients of 210 to 273 gY_p/kg MP for the USA to Nordic system imply that these cattle were within the curvilinear, diminishing phase of the response to increasing MP wherein the efficiency of the marginal response to increasing MP was 0.21 to 0.27.

With this in mind, the data from NIRD, GRI and Boxworth used in the IDWP (1991) trial can be re-examined, together with data from Hillsborough (Mayne and Gordon, 1985) which also contains all the information necessary to calculate MP supply according to IDWP and relate it to Y_p. The results of this exercise are illustrated in Figure 5.2 and appear to bear a close similarity to those reviewed by Van Straalen (1991). Within each centre there is a clear and close association between MP supply and Y_p although the intercepts differ markedly between centres.

Farms a, b, c and d are Hillsborough, GRI, NIRD and Boxworth respectively. Individual regression equations for the four farms were as follows.

		s.e.b.	r^2
Hillsborough	Yp = 309 + 0.176MP	0.051	0.70
GRI	Yp = 286 + 0.291MP	0.057	0.87
NIRD	Yp = 539 + 0.137	0.944	0.55
Boxworth	Yp = 489 + 0.221	0.054	0.70

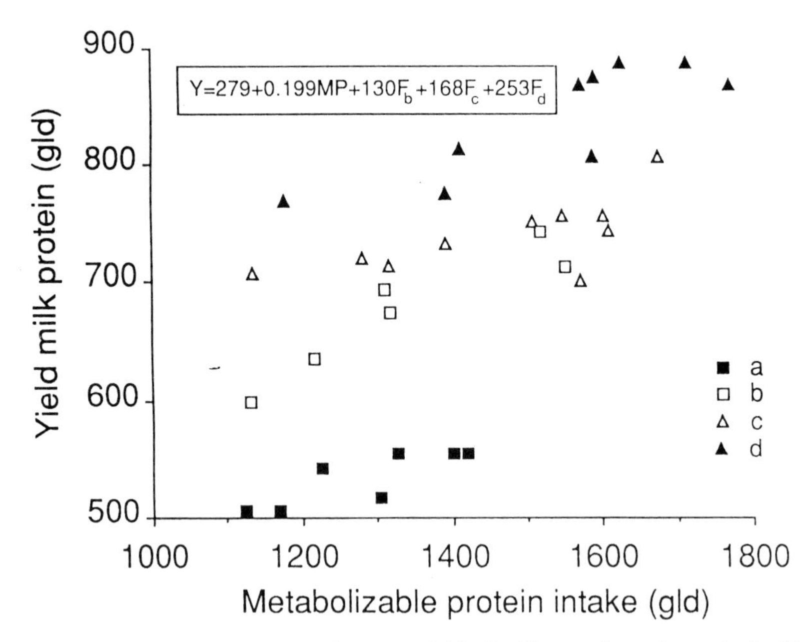

Figure 5.2. The relationship between yield of milk protein and metabolizable protein intake in trials at four sites; a, Hillsborough; b, G.R.I.; C, N.I.R.D.; and D, Boxworth

The overall regression equation becomes $Y_p = 279 + 0.199MP + 130F_b + 168F_c + 253F_d$ $r^2 = 0.94$. The implication of these observations is that Y_p does correlate closely with protein supply as defined by MP but in practical circumstances the marginal response to increasing MP is far below any of the conventional estimates of k_{nl}, made in circumstances where MP is obviously limiting Y_p. In the practical circumstances illustrated by Figure 5.2, the marginal response to increasing MP is 0.2 (to 1 significant figure). In practice it does not matter too much whether this response is due to supply of limiting amino acids, substrates for gluconeogenesis, or a stimulation of appetite for forage. It is a production response which has been described with some precision by the MP system.

How is it possible to reconcile this marginal efficiency of 0.20 for the response to increments of MP with the IDWP value of 0.68 for the net efficiency of utilisation of MP for lactation? Figure 5.3 attempts to resolve this apparent paradox. The slope defined by the solid line k_{nl} relates Y_p to MP for a 600kg cow when maintenance requirement for MP = 310g/d and $k_{nl} = 0.68$. The four lines a, b, c and d are the regression lines summarising the results from the four farms illustrated more fully in Figure 5.2. It now becomes clear that k_{nl} defines with some accuracy (if not precision) the relationship between mean MP supply and mean Y_p on each farm. This cannot however be taken as evidence that the net biochemical efficiency of MP for milk synthesis is approximately 0.68 since it is impossible to consider response to MP in isolation from response to ME.

The interrupted line, k_l in Figure 5.3 relates Y_p to MP on the basis that the increase in milk yield is a response to energy not protein. The assumptions used to derive this slope are as follows: Milk contains 32g protein/kg, ME requirement = 5 MJ/kg milk, thus Y_p/ME = 6.4g/MJME. The optimal ratio MP:ME is 9.2g/MJ (i.e. 1838/199 from

Table 5.2, when CP = 173g/kg). When CP = 160g/kg. the ratio MP:ME = 8.5. Applying the same logic in reverse one can calculate the crude protein concentration in the diet that would predict a slope of $k_{nl} = 0.68$ if the response in milk yield were attributable to ME not MP. It becomes 176g/kg, effectively that predicted as optimal in Table 5.2.

These calculations suggest that $k_{nl} = 0.68$ may be no more than the coefficient that relates Y_p, coincidentally to MP when milk yield is primarily determined by ME intake. The data of IDWP (1991) and Mayne and Gordon (1985) presented in Figures 5.2 and 5.3 do, in fact, provide further evidence for this assumption. Figure 5.2 reveals a good consistency between trials in the marginal efficiency (0.20) of response to MP but does not account for the large differences in intercept. It is not possible to calculate exact ME intakes for the four farms but using values for DM intake one may conclude that ME intakes were very similar within farms (except for control diets) but differed substantially between farms. Estimated mean ME intakes on farms a, b, c and d are 142, 174, 168 and 190 MJ/day respectively. The closed circles on each line in Figure 5.3 relate milk yield in each case to ME intake at the optimal ratio MP:ME. Thus the difference between farms in milk yield can be attributed primarily to supply of ME not MP.

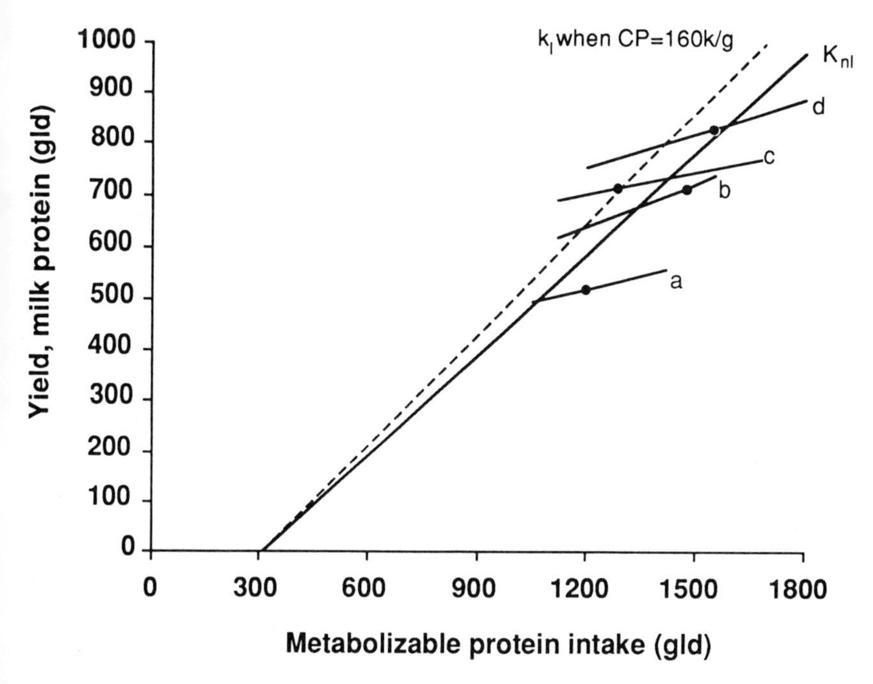

Figure 5.3. Observed and theoretical relationships between metabolizable protein intake and yield of milk protein. The four lines are the observed regressions at farms a, b, c and d. The solid line k_{nl} is the predicted response to MP when $k_{nl} = 0.68$. The interrupted line k_l relates Y_p to MP when CP concentration = 160g/kg on the assumption that milk yield is determined by metabolizable energy intake. (For further explanation, see text)

It is possible to pursue too far the argument as to whether k_l determines k_{nl} or vice versa. ME and MP supply are obviously autocorrelated and the coefficients $k_l = 0.62$ and $k_{nl} = 0.68$ appear to define the relationship between yield of milk energy and protein and intakes of ME and MP above maintenance when MP and ME are in balance. It is suggested therefore that the metabolizable protein system, including the coefficient k_{nl} = 0.68 can be used to calculate MP requirement for a stated milk yield (say 30kg/day) in the first instance and will ensure a physiologically correct balance between MP and ME. If then one wishes to manipulate milk yield, e.g. to match quota and ME intake is already constrained by the limit of gut fill, then it is possible to vary the ratio MP:ME on the assumption that the marginal efficiency of response to MP alone will be 0.20.

Conclusions

There can no longer be any doubt that DCP does not constitute a sound basis for feeding ruminants. The UK has lagged behind the rest of Europe and the USA in deciding upon a new system which distinguishes between ruminal and post-ruminal digestion. Thanks to this delay IDWP has managed (1) to avoid the systematic errors of prediction in the ARC (1980, 1984) systems; (2) to incorporate elements of flexibility in, e.g. FME:ME, ERDP, dUP, not present in the first of the new protein systems.

As it stands, the IDWP report is less definitive than the committee would have liked. The estimation of FME, ERDP, DUP and the yield and efficiency of MCP remains imprecise but these things can be improved without any conceptual change to the system. There is every reason to conclude that MP constitutes a satisfactory and inherently improvable basis for the description of nutrient supply. All the new protein systems get into difficulties both numerical and conceptual when attempting to relate supply to requirement. I have wrestled for many years with problems of belief relating to the sacred dogma of protein nutrition and now feel able to speak two great heresies:

1. Measurements of the efficiency of utilisation of amino acids based only on the entry of amino acid N to the duodenum are, in ruminants, an expensive irrelevance since they define neither true supply of proteins and NPN (i.e. through the mouth) nor the loss of efficiency (as defined) implicit in the recycling of endogenous N.

2. Classic measurements of the linear response to increments of protein in an individual amino acid made inevitably in circumstances where that protein or amino acid is severely limiting production are necessary for evaluation of feed proteins but do not define marginal efficiencies of response in commercially valid production trials.

The IDWP report as it currently stands does not inspire confidence that the MP system has been adequately tested even with dairy cattle for which it is most important. It is suggested however that this is due less to a lack of good data than to a failure to abandon dogma such as NAN flow at the duodenum and to confusion between production response and biochemical efficiency. The Metabolizable Protein System as proposed can be used to predict MP requirement for a stated level of milk production, in the first instance, and to achieve a proper balance between ERDP, DUP and ME in the diet. For this purpose the efficiency term k_{nl} is satisfactory. Thereafter the marginal efficiency of response to increments of MP when ME is fixed is approximately 0.20. This information is not by itself sufficient to define efficiency for an all-embracing model of the nutritional and genetic determinants of milk yield such as that proposed by Oldham and Evans (1989) but it does, in practice, permit fine tuning of ration formulation about the status quo to meet more immediate demands, such as the need to achieve milk quota at least cost.

References

Agricultural Research Council (1980). *The Nutrient Requirements of Farm Livestock, No.2. Ruminants. 2nd edn.* Commonwealth Agricultural Bureaux, Farnham Royal

Agricultural Research Council (1984). *The Nutrient Requirements of Ruminant Livestock. Supplement No.1.* Commonwealth Agriculture Bureaux, Slough

Alderman, G. (1987). In *Protein Evaluation of Ruminant Feeds.* pp. 283–297. Eds. Alderman, G. and Jarrige, R. pp. 283–297. Commission of European Communities

Chaudhry, A.S., Webster, A.J.F. and Marsden S. (1992). *Animal Production* (in press)

Chen, X.B., Hovell, F.D.de B., Orskov, E.R. and Brown, D.S. (1990). *British Journal of Nutrition,* **63**, 131–142

Dewhurst, R.J. and Webster, A.J.F. (1992). *British Journal of Nutrition.* (in press)

Interdepartmental Working Party (1991). *Report on the Protein Requirements of Ruminants.* Nutrition Abstracts and Reviews (in press)

Institut Nationale Recherche Agronomique (INRA, 1988). *Alimentation des Bovins, Ovins et Caprins.* pp. 370. Ed. Jarrige, R. INRA, Paris

Klopferstein, T. and Britton, R. (1987). *Proceedings of Distillery Feed Conference USA,* **42**, 84–86

Madsen, J. (1985). *Acta Agricultura Scandinavia, Suppl.* **25**, 9–20

Mayne, C.S. and Gordon, F.J. (1985). *Animal Production,* **41**, 269–280

National Research Council (1985). *Ruminant Nitrogen Usage.* U.S. Academy of Science, Washington, D.C.

Oldham, J.D. (1987). In *Protein Evaluation of Ruminant Feeds.* Eds. Alderman, G. and Jarrige, R. Commission of European Communities.

Oldham, J.D. and Emmans, G.C. (1989). *Journal of Dairy Science,* **72**, 3212–3229

Orskov, E.R. and McDonald, I. (1979). *Journal of Agricultural Science, Cambridge,* **92**, 499–523

Owens, F.N. (1982). *Protein Requirements for Cattle.* Oklahoma State University, Stillwater

Rohr, K. (1987). In *Feed evaluation and protein requirement systems for ruminants.* pp. 3–10. Eds. Alderman, G. and Jarrige, R. Commission of European Communities

Tamminga, S. (1982). In *Protein contribution of feedstuffs for ruminants.* pp. 4–17. Ed . Miller, E.L., Pike, I.H. and Vares, A.J.H. Butterworths, London

Van Soest, P.J. and Mason, V.C. (1991). *Animal Feed Sciences and Technology, 32*, 45–53

Van Straalen, W.M. (1991). The validation of protein evaluation systems by means of milk production experiments with dairy cows.

Verite, R., Michalet-Doreau, B, Chapoutot, P., Peyraud, J.L. and Poucet, C. (1987). *Bulletin Technical C.R.Z.V. Theix.*, INRA **(70)**, 19–34

Waters, C.J. (1991). *Protein Quality of Feeds for Ruminants.* Ph.D. Thesis, University of Bristol

Waters, C.J., Kitcherside, M.A. and Webster, A.J.F. (1992). *Animal Feed Science and Technology* (in press).

Webster, A.J.F. (1987). In *Protein evaluation of ruminant feeds.* pp. 47–54. Eds. Alderman, G. and Jarrige, R. Commission of European Communities.

Webster, A.J.F., Dewhurst, R.J. and Waters, C.J. (1988). In *Recent Advances in Animal Nutrition.* pp. 167–191. Eds. Haresign, W. and Cole, D.J.A. Butterworths, London

III

General Nutrition

6

IMPLICATIONS OF NUTRITION OF ANIMALS ON ENVIRONMENTAL POLLUTION

S. TAMMINGA and M.W.A. VERSTEGEN

Department of Animal Nutrition, Agricultural University, Wageningen, The Netherlands

Introduction

Present intensive animal production systems are inefficient converters of feed into desired animal products. This is particularly true for Nitrogen (N), Phosphorus (P) and Potassium (K). A large fraction of these elements in feed is not deposited in desired animal products, but wasted in animal manure as a mixture of faeces and urine. Losses in animal excreta are as a mixture of solids, aqueous solutions and gases.

Because of the high animal numbers per unit area of land, resulting in a high ratio between waste and land area, animal manure is becoming an increasing burden on the environment. This is particularly the case in areas where the animal industry has developed intensive systems, such as in many areas of The Netherlands. Public pressure aimed at reducing environmental pollution, including that caused by the animal industry, is growing. In order to avoid a forced, significant reduction in the size of the animal industry, measures will have to be taken to reduce its negative impact on the environment.

The animal industry in The Netherlands is characterized by largely being based on imported feeds. This is particularly the case with pig and poultry production. Total feed consumption by livestock is some 25 million tonnes of dry matter, 14 million tonnes of which is as concentrates. Over 80% of the ingredients used in concentrates are imported. Some 70% of the Dutch animal products are exported and as a result, the Netherlands' animal industry causes the deposition of enormous amounts of undigested and/or non-utilized feed residues through faecal and urinary excretion. This is particularly true for N and to a lesser extent for P. Simultaneously, considerable amounts of carbon dioxide (CO_2), methane (CH_4), ammonia (NH_3) and other volatiles are produced and lost to the atmosphere.

Until now, nutritional management of farm animals has been used as a tool to maximize the output of useful products like milk and meat with little or no attention being paid to the output of less desirable products like faecal and urinary excretion of N and P or the loss of CO_2, CH_4, NH_3 and other volatiles. Nutritional management can also be used as a tool to help control environmental pollution. The objective of this paper is to present the state of the art on the situation in The Netherlands and to discuss measures introduced so far and plans which will be implemented in the near future, with emphasis on nutritional aspects.

Nature of N losses in farm animals

Losses of nitrogen (N) and ammonia (NH3) in different classes of livestock in The Netherlands are summarized in Table 6.1 (Coppoolse *et al.*, 1990; Van der Meer, 1991).

Table 6.1. ESTIMATED ANNUAL INTAKE, EXCRETION AND LOSSES OF N AND NH_3 BY ANIMAL PRODUCTION IN THE NETHERLANDS IN 1986

Category	*Numbers present* (m)	*N input*	*N excretion*	*NH_3 - N loss storage*	*spreading*
			——— '000 t ———		
Lactating cows	2.3	417	357	43.5	50.2
Young cattle	1.7	125	117	14.2	16.4
Beef cattle	0.4	19	18	2.2	2.5
Veal calves	0.7	17	10	1.3	1.3
Sheep	1.0	6	5	0.5	0.5
Fattening pigs	6.9	138	106	20.6	17.2
Piglets	4.8	24	15	3.0	2.5
Young sows	0.4	8	7	1.2	1.0
Breeding sows	1.3	37	31	4.7	5.2
Laying hens	14.3	33	25	2.5	2.9
Broilers	44.0	76	53	4.8	6.3
Total	77.7	900	734	98	106

Sources: Copoolse *et al.* (1990); Van der Meer (1991)

Sources which contribute to faecal N are undigested feed N, microbial N and endogenous N. Undigested feed N and endogenous N are mainly in true protein (amino acids), whereas microbial N is partly (15-20%) present in nucleic acids. In all species of farm animals microbial degradation of protein in the hindgut may mean that faecal N is partly excreted as NH_3.

Urinary N is mainly excreted in urea in cattle, sheep and pigs and in uric acid in poultry. Other N containing compounds excreted in urine of all species are allantoin, hippuric acid and creatinine. Urea is excreted by the kidneys but originates in the liver as the end product of the detoxification of NH_3.

After excretion, faeces and urine of farm animals usually get mixed. Freshly voided faeces and urine already contain volatile components, but only low levels. Animal manure contains soluble and easily degradable organic matter, including urea and easily degradable protein. Due to microbial activity in the manure, extensive degradation of N containing compounds develops (Table 6.2). This causes a rapid breakdown and the formation of (partly branched chained) volatile fatty acids (VFA) and other degradation products of protein (Spoelstra, 1978). Because of the presence of ureolytic bacteria in faeces and large amounts of urea in urine, NH_3 is produced very rapidly during storage. Between 60 and 75% of the N in excreted manure is converted into NH_3 (Klarenbeek and Bruins, 1988), of which between 25 and 40% is lost during storage and an additional 20 to 60% during spreading.

Table 6.2. N LOSSES FROM ANIMAL WASTE DURING STORAGE

	Cattle	*Pigs*	*Poultry*	*Total*
As excreted[a]				
N ('000 t)	507	159	68	734
P ('000 t)	62	37	18	117
Ratio N/P	8.2	4.3	3.8	6.3
After storage[b]				
N ('000 t)	341	134	54	529
P ('000 t)	62	37	18	117
Ratio N/P	5.5	3.6	3.0	4.5
N "loss" ('000 t)	166	25	14	205
(%)	33	16	20	28

[a] Van der Meer, (1991). [b] Anonymous, (1988a).

Nitrification and denitrification convert an unknown proportion of the NH_3 further into nitrate and gaseous N_2. As a side effect oxides like nitrous oxide (N_2O), nitrogen oxide (NO) and nitrogen dioxide (NO_2) are produced and also escape into the atmosphere. These gaseous intermediates, N_2O in particular, are believed to contribute to the impairment of the ozone layer and are therefore considered undesirable. Total loss of N_2O in agriculture in The Netherlands was estimated at between 10 and 15 million kg per year (Goossensen and Meeuwissen, 1990).

Animal waste contains over 60 volatile components (Spoelstra, 1978). Of these about twelve are considered important as contributors to malodour. Noxious odours in animal manure can be classified (Coleman *et al.*, 1991) as carboxylic acids (acetic acid, butyric acid), phenolics (p-cresol, phenol), aliphatic and nitrogen containing compounds (ammonia) or sulphur containing compounds (hydrogen sulphide, demethyl sulphide, ethyl mercaptan, methyl mercaptan). Table 6.3 gives a summary of the most important characteristics of these volatile components.

Table 6.3. CHARACTERISTICS OF VOLATILES IN ANIMAL WASTE

Component	*Form*	*Smell*	*Target organs*	*TLV* (ppm)
Acetic acid	colourless liquid	pungent acrid	Respiratory System (R.S.)	10
Butyric acid	colourless oily liquid	pungent	skin, eyes	
Propionic acid	colourless oily liquid	rancid	R.S., eyes	
Phenol	colourless to pink crystals	odour	R.S., eyes	5
p-Cresol	colourless to pink crystals	carbolic odour	R.S., eyes, skin liver, kidneys	5
Ammonia	colourless gas	pungent	R.S., eyes	
Nitrogen dioxide	red brown gas	toxic	R.S., lungs, eyes, skin	5
Nitrogen oxide				25
Ethyl mercaptan	colourless liquid	garlic odour	R.S., eyes, mucous membranes	0.5
Methyl mercaptan	colourles gas	nauseous smell		
Hydrogen disulphide	colourless gas	offensive odour	R.S., eyes	10

For reasons of environmental protection, at present most attention is on NH_3. Wild plants appear to be extremely sensitive to NH_3 and recommended maximum levels range from 30 to 40 ppb for "wild" vegetation on poor soils to around 100 ppb for cultivated plants (Anonymous, 1988). Humans and farm animals are less sensitive to NH_3. The Threshold Limit Value (TLV) for humans at a daily subjection time of 8 hours is 25 ppm. At high levels NH_3 may be harmfull to the respiratory organs of farm animals and because farm animals are subjected to NH_3 almost continuously, lower threshold levels of between 2 and 10 ppm are considered safe.

The potential health risk of volatiles in animal manure is only one aspect. An additonal problem is their bad smell. In a review on gaseous pollutants from animals (Tamminga, 1992), it was concluded that in general their Odour Threshold Values (OTV's) are at least a factor 500 lower than TLV's and hence, these odours can easily be recognized, long before their concentrations become a health risk. "Normal" values in farm buildings (pig houses) are usually far below the TLV, except NH_3 and H_2S, which may occasionally reach dangerous levels.

Losses of other gaseous components

An important gaseous component lost by all species of farm animals is carbon dioxide (CO_2). In quantitative terms annual production of CO_2 is in the order of 4000 kg per mature bovine and 400 kg per mature ovine. In comparison, a human produces some 300 kg of CO_2 from metabolism per year and an average passenger car, using 1750 kg (2500 kg) of fuel, 5500 kg (Tamminga, 1991). Farm animals, which have a metabolic rate about 50-80% higher than that of humans, produce relatively more CO_2. A pig of 50 kg may produce about 450 kg per year. CO_2 is held partly responsible for the greenhouse effect, believed to cause global warming. Because farm animals produce CO_2 from renewable and not from fossil energy, it is considered harmless and will not be discussed further.

Ruminants also produce methane (CH_4) at a rate of between 2 and 12% of their gross energy (GE) intake (Johnson *et al.*, 1991), which causes concern because its contribution per mole to the greenhouse effect is estimated to be 15 times that of CO_2. CH_4 is also considered to contribute to the impairment of the ozone layer. It is a colourless non-toxic gas, but in the absence of oxygen it can have narcotic effects when present in high concentrations. In high concentrations it is highly inflammable and therefore risky. At a daily production rate of 425 l of CH_4, a total of 110 kg is produced annually per mature bovine. Pigs produce small amounts of CH_4 too, but this is less than 1% of GE intake and compared to the production of ruminants insignificant, slightly over 1 kg of CH_4 per animal per year.

Animal waste as a raw material

Recycling of animal waste as organic fertilizer depends on the availability of sufficient spreading surface. Cultivated land in The Netherlands amounts to 1.91 million ha of which 1.14 million ha is used as permanent grassland, 0.20 million ha for growing corn silage and 0.57 million ha for arable farming and horticulture.

N excreted in animal waste (734 million tonnes, Table 6.1) is potentially sufficient for a fertilization rate of over 350 kg N/ha. In addition some 500 million tonnes of N is applied as inorganic fertilizer, bringing the total potential input to 850 kg/ha. Of the N in organic waste some 200 million tonnes is believed to be volatilized and lost, of which 50 million tonnes returns as NH_3 (Van der Meer, 1991). This brings the total input to 700 kg N/ha. No doubt exists that flow of N in agriculture in The Netherlands is out of balance. Several schemes are under development to regain some sort of a balance again.

The first option is to reduce volatilization and loss of NH_3. Several ways which will prevent NH_3 to escape from animal waste during storage are under development

(Tamminga and Verstegen, 1991). At present most emphasis is on airtight covering of the storage pit and acidification of the manure. The latter seems promising, because animal manure becomes enriched with N and is therefore more valuable as N fertilizer (Esteban Turzo *et al.*, 1988; Pain *et al.*, 1990). Denitrification may however convert part of the trapped N into N_2, which escapes in the air and gets lost.

A further important loss of NH_3 is during spreading. According to Rainelli (1989) losses of NH_3 in The Netherlands are distributed as in Table 6.4. In order to reduce NH_3 losses, recommendations have also been made and some of them will become enforced by law within the near future. They include the application of manure during the growing season only and application by soil injection. Clearly these attempts will only be succesful if a significant part of the inorganic manure is replaced by organic manure.

Table 6.4. SOURCES OF AMMONIA EMMISSION IN THE NETHERLANDS

Animal species	*Barn and storage*	*Spreading*	*Pasture*	*Total*
Cattle	17%	32%	10%	59%
Pigs	11%	18%	-	29%
Poultry	7%	5%	-	12%
Total	35%	55%	10%	100%

From Rainhill (1989)

An alternative which is being extensively explored is the upgrading of animal manure to a dry fertilizer, so that it becomes a tradable commodity. Progress in this area is slow, not only because of technical difficulties, but also because the success of such "manure refineries" largely depends on a suitable and large enough market and prospects for that are uncertain.

Use as a fuel is limited because of a high water content. There is evidence that by reducing protein intake, which is possible if protein quality is improved, the water intake and as a consequence water excretion can be reduced (Pfeiffer and Henkel, 1991). Some application is also possible after conversion into CH_4 (Zeeman, 1991). Recycling of animal manure to animal feed has only limited prospects because of its low nutritive value and possible health risks. The best prospects in this area are with poultry droppings.

Reducing N losses in animal waste through feeding management

The measures discussed above are not the only ones that can help to bring N flow in agriculture back into balance. Feeding management is also an important tool to reduce N excretion in animal waste, in monogastric animals (Lenis, 1989) as well as in ruminants (Tamminga, 1991).

Two approaches are possible, enhancement of the deposition of N in animal products (meat, eggs, milk) while N input is maintained constant or reduction of dietary N input while productivity is maintained constant. The first approach requires the intermediary metabolism to operate more efficiently. The succes of the second approach largely depends on reducing N losses along the gastro-intestinal tract.

Both approaches will result in a reduced excretion of N in animal manure. A more efficient intermediary metabolism will reduce N excretion in urine, whereas a reduction in losses from the gastro-intestinal tract will reduce excretion both in faeces and in urine. Although feeding management can influence the composition of animal manure when voided, little information is available on its effects on manure during storage. It can be expected that short term volatilization of NH_3 from pig manure can be reduced by altering the ratio of N losses between faeces and urine. The effect of long term effects is more difficult to predict. Hence, improved feeding management is no guarantee that the quality of manure after storage will be significantly improved.

N losses from the rumen and ways of reducing them

In ruminants, the rumen is an important site of N losses. Estimates suggest (Tamminga, 1991) that for instance in a dairy cow between 25 and 50 kg of N is annually lost from the rumen. N losses from the rumen result from a surplus of rumen degradable N (RDN), a lack of synchronisation between rate of degradation of rumen degradable protein and rumen degradable carbohydrates or a less efficient capture of RDN by rumen microbes. An easy and quick recognition of ruminal N losses is achieved with protein evaluation systems in which the balance between rumen degraded N and microbial N capture is indicated, as is the case in the Scandinavian (Madsen, 1985) and the Dutch protein evaluation systems (Anonymous, 1991).

In The Netherlands, ruminal N losses in the grazing animal are amongst the highest. Due to intensive grassland management, including the application of high levels of N fertilizer, grass with a (crude) protein content of up to 250 g/kg of DM is quite common. Not only is this level of crude protein high, rate of ruminal degradation is high too, resulting in effective degradations of 70 to 80% (Van Vuuren *et al.*, 1986). Only part of the RDN is captured by rumen microbes and as a consequence the result is extremely high N losses. In fresh grass a high N level is almost a guarantee for a high nutritive value, but the N level is about twice as much as the grazing animal needs and the amount which it can efficiently utilize. High levels of fertilizer N not only result in excessive levels of N in grass, but will also almost completely eliminate leguminous forages like clover, which have a capacity to bind N through N-fixation.

Vertregt and Rutgers (1988) estimated that in grazing animals 15% of the urinary N is lost as NH_3, and an even larger proportion was not recovered from the soil or the grazed herbage. This suggests that there are other important losses too. High urinary excretions lead to concentrations of N in the soil at sites where animals urinate, equivalent to a once-only application of 700 kg N/ha, with the danger of high nitrate levels.

Improvements can be achieved by reducing the high level of N fertilization, which at present is at a rate of over 300 kg of N/ha. Lower N fertilization not only results in a lower N level in the grass (Table 6.5), but also reduces dry matter yield. Increasing the growing days may overcome this, but will reduce its nutritive value and as a consequence reduce grass intake. As a consequence of the milk quota system most dairy farms already have a surplus of forage, hence a lower productivity seems possible and justified. Alternative options which will reduce the undesirable large urinary N excretions are supplementation of grass with low N forages or concentrates, such as maize silage, beet pulp or cereals. The deficit of N in these supplements can easily be balanced by the surplus in grass resulting in a considerably improved overall N utilization (Valk *et al.*, 1990).

Table 6.5. EFFECT OF GROWING DAYS AND LEVEL OF N FERTILIZATION ON CRUDE PROTEIN CONTENT (%) OF FRESH GRASS

Growing	kgN/ha/yr						
days	0	250	275	400	500	550	700
7	21.2	27.5	-	28.8	-	30.0	33.8
13	-	-	-	-	31.4	-	-
23	-	-	27.8	-	33.9	-	-
29	-	-	26.9	23.1	-	27.5	28.8
42	-	17.5	-	-	-	22.1	-
60	9.4	-	-	-	-	20.0	-

Source: Tamminga and Van Vuuren (unpublished)

Lower dietary N levels, a reduced degradation of feed protein or a more efficient capture of rumen degraded N (RDN) seem also promising ways to reduce N losses from the rumen (Tamminga, 1991). These measures are easier to realize with concentrates than with forages.

A lower N intake can easily be achieved by the inclusion of cereals in concentrates although at present this is not economically attractive. N content of the total diet should be kept at a minimum of 24 g/kg of DM in order to avoid impaired rumen digestion. The replacement of grain by-products, which at present form the main body of dairy concentrates in The Netherlands, by cereals, may create new environmental problems, because some by-products can only economically be used in the wet form in ruminant feeding. This is particularly true for a number of wet by-products like brewer's grains, pressed beetpulp, corn gluten feed and vinasse.

A reduced rate of degradation is possible by selecting appropriate feed ingredients or by processing, such as treatment with heat or with chemicals (Van Straalen and Tamminga, 1990).

N losses from the rumen can also be controlled by feeding method, e.g. feeding frequently or feeding totally mixed rations (TMR). Such feeding practices will synchronize the availability of rumen degradable N and rumen degradable carbohydrates (Tamminga *et al.*, 1990a) and therefore stimulate the efficiency of microbial N capture. Finally, the amount of ensiled products in ruminant diets should be controlled, because they contain significant amounts of fermentation end products which are poorly used as an energy source by rumen microbes. These can be minimized by pre-wilting before ensiling.

N losses from the intestine and ways of reducing them

Some N losses from the small intestine occur because of incomplete digestion of feed N. In monogastric animals protein digestibility is low for some legume seeds, due to the presence of Antinutritional Factors (ANF) like lectins or protease inhibitors (Huisman, 1990). Low protein digestibility can be overcome by technological treatment, in an optimal combination of temperature, moisture and time. Short treatment at high temperature is more effective than a lower temperature for a longer period (Table 6.6). Heat treatment is also effective in reducing the ANF content or activity mentioned earlier (Van der Poel *et al.*, 1991). Other potential ways to reduce the activity of proteinous ANF (lectins, protease inhibitors) or the excretion of endogenous protein caused by NDF and other NSP's are the application of enzymes, germination (Tamminga *et al.*, 1990b) or grinding to a finer particle size (Wunsche *et al.*, 1987).

Table 6.6. EFFECT OF TECHNOLOGICAL PROCESSING OF BEANS ON PROTEIN DIGESTIBILITY (%) IN PIGS

Time	*Temperature*° C		
(minutes)	102	119	136
1.5			80.9
5		76.9	
20	69.2		
40	73.5		
60	74.5		
80	76.0		

Van der Poel (1990)

A second, but at least equally important source of N losses from the small intestine is endogenous secretions. Endogenous protein is secreted into the intestinal lumen as enzymes, mucus and epithelial cells (Swanson, 1982). It is mixed with the digesta when passing along the digestive tract and subjected to proteolytic digestive enzymes. Digestion and reabsorption of endogenous protein are however incomplete and therefore endogenous protein contributes to N in animal waste, either directly or after microbial conversion in the hindgut. In the latter case it may partially be excreted in urine. After microbial degradation, followed by absorption of the resulting ammonia, NH_3 is converted to urea in the liver and excreted in the urine by the kidneys. Endogenous protein has to be resynthesized at the expense of considerably more amino acids than are incorporated into newly synthesized protein (Simon, 1989). In sheep endogenous protein was shown to be related to passage of dry matter along the small intestine (Van Bruchem *et al.*, 1989). Additional sources of endogenous protein in monogastric animals are ANF and possibly NSP.

Potentially, endogenous losses can be reduced by an improved dry matter digestibility. Technological treatment (heat, moisture, pressure) also appears to be a promising way to reduce excessive endogenous losses (Tamminga and Verstegen, 1991). The effect of both measures has not yet been thoroughly researched.

N losses in intermediary metabolism and ways of reducing them

Protein requirements are defined as the amount of protein above which extra protein does not result in extra growth or protein deposition (Fuller, 1991). This means that in practice protein is supplied to a point where the marginal efficiency of utilization has become zero. In the light of the need for an improved N economy of animal production systems this approach can be challenged. Environmental restrictions may force future animal production systems more to be based on optimal or marginal dose- response relationships than on meeting requirements.

Protein deposition in animals requires essential amino acids to be supplied in a well balanced and strictly defined ratio. For monogastric animals the concept of "ideal" protein was developed (ARC, 1981), based on the amino acid composition of whole body protein. The basis for such an ideal protein may only partly be valid, because it does not take into account losses in maintenance (Moughan and Verstegen, 1988) and the amino acid pattern required to maintain N equilibrium and that required to deposit protein are quite different (Fuller *et al.*, 1989). The required amino acid ratio almost certainly varies also among the different organs and tissues. Besides, the efficiency of utilization of amino acids for protein deposition varies between amino acids. Fuller (1991) estimated the lowest value (0.64) for tryptophane and the highest value (0.93) for lysine, even when "ideal" protein was fed. Little information is available on changes in the marginal response for individual amino acids. In veal calves it was shown that changes in the marginal response were much larger for methionine + cystine than for lysine (Van Weerden, 1989).

Whether dose-response curves between absorbed protein and protein deposited show a linear-plateau (Campbell, 1988) or a curvilinear (Van Weerden, 1989) relationship is still a matter of some debate. In both concepts an animal has a maximum capacity to deposit protein depending on age, liveweight and physiological state, regardless of the supply of energy and/or protein. To what extent the maximum deposition is achieved depends on the ratio in which energy and protein are supplied and on the amino acid composition of the protein.

Contrary to what most newly developed protein evaluation systems for ruminants would suggest by using a constant conversion of absorbed protein to milk protein, responses in milk protein yield of lactating dairy cows to an increased supply of protein also have a diminishing efficiency (MacRae *et al.*, 1988; Verité and Peyraud, 1990; Tamminga, 1991), similar to what is usually observed in growing monogastric animals (Fuller, 1991). This also explains why an increased supply of rumen protected protein often has a variable and rarely a spectacular effect and why extra casein infused in abomasum or proximal duodenum is utilized for milk protein production with such a low marginal efficiency (MacRae *et al.*, 1988).

Nutrients supplied to farm animals are usually partitioned between fat and protein, a process which is largely controlled by the animal's endocrine balance. Endocrine balance is influenced by its physiological status (Bauman, 1984), plane of nutrition and environmental temperature (Ingram and Dauncey, 1986).

Protein requirements vary with age and physiological status. In a growing animal the ratio between protein and fat deposition changes and in all species of farm animals the ratio between requirements for maintenance and for production changes. As a consequence amino acid requirements change, not only in comparison with energy requirements, but the ratio in which amino acids are required may also change. In order to keep unnecessary N losses to a minimum, amino acid supply has to change almost continuously, which can be achieved by phase feeding.

In phase feeding, animals are fed according to the requirements of their physiological status and age. Replacing the two phase system presently used in fattening pigs, by a three phase system would reduce N excretion by some 6% (Lenis, 1989). Replacing a one phase feeding system in pregnant and lactating sows by a two phase system reduced N excretion in manure by 20-25% (Everts and Dekker, 1991). A disadvantage of phase feeding is that it requires investment in equipment which can mix two or more feeds. Phase feeding should not be restricted to growing monogastric animals. A suitable application of this practice in ruminants is feeding totally mixed rations (Tamminga, 1991).

Losses of N in intermediary metabolism are largely related to protein turnover, which varies from less than 10% to over 60% per day between different organs and tissues (Simon, 1989). High turnovers are found in digestive tissue, pancreas and liver. In skin and muscle turnover is much lower, but, because of its large pool size, turnover in muscle still makes a large contribution to the total N loss in intermediary metabolism. The dynamic state of continuous turnover not only causes large energy losses (Van Es, 1980), but the re-utilization of amino acids takes place with an efficiency of considerably less than 100%. Estimates vary between 75 and 95% (Simon, 1989). Losses in turnover vary among organs and tissues and with nutritional status. Increasing the protein supply increases turnover (Simon, 1989). Part of the N losses are in maintenance, a significant part of which is believed to be lost in the resynthesis of endogenous losses. Excretion is as endogenous faecal losses as well as urinary losses of urea.

Control of turnover is as yet poorly understood, but the balance between supplied nutrients as well as the endocrine balance is important. The application of pST considerably improved N deposition in growing pigs. Marginal response to protein input was also increased from about 20% in the control groups to over 45% in the pST treated groups

(Table 6.7). It was calculated that the application of pST could result in a reduction of urinary N excretion of around 20% (Van Weerden, 1989).

Nutrient repartitioners (hormones, beta-agonists) seem powerful tools to optimise the endocrine balance, and to reduce waste of N. Lack of consumer acceptance and hesitation among politicians will most likely prevent their introduction (Lamming, 1986). The introduction of transgenic animals, which could act in a similar direction, will probably face even more difficulties.

Plant proteins rarely supply amino acids in the required ratio. The ratio of the mixture of amino acids supplied to organs and tissues can be improved by mixing plant proteins of different amino acid composition. In pigs and poultry scope exists for using synthetic amino acids (Lenis, 1989; Schutte, 1989). At present this is rarely economically feasible. New biotechnological production techniques may change this. Availability at competitive prices is presently restricted to lysine and methionine. If in future they become available at competitive prices, there is scope for threonine and tryptophane to be used additionally (Lenis, 1989).

At present non-essential amino acids have to be supplied in excess to the animal to ensure an adequate supply of essential amino acids. If essential amino acids in addition to the ones already discussed become also available in large quantities and at competitive prices this surplus of non-essential amino acids can be reduced (Lenis, 1989).

Synthetic amino acids do not need hydrolysis prior to absorption and their absorption may therefore be more rapid than amino acids supplied as protein. To exploit the full potential of synthetic amino acids, more frequent feeding may be required, particularly in poultry. Provided they are protected against degradation in the rumen, there may even be scope for using synthetic amino acids in diets for dairy cows.

An equally promising way in dairy cows is to prevent reduction in milk protein content. Feeding conditions causing a reduced milk protein content are underfeeding, a common problem in early lactation, replacing non-structural (concentrates) by structural (forage) carbohydrates, including extra lipids or large amounts of silage or other prefermented ingredients in the diet (Tamminga, 1991). Such factors are easily avoided.

A long term prospect is the development of integrated dynamic feed evaluation systems, which will predict the response to a combination of nutrients supplied to an animal in a given physiological condition and a given environment as outlined by AFRC (1991). In future such integrated systens have to replace the presently used factorial systems. The present systems are not only factorial but also separated in systems for energy and for protein. In integrated feed evaluation systems characteristics of animals and feeds and interactions between feed ingredients as well as between animal and feeds can be incorporated. The modelling of growth in pigs, and of rumen fermentation in cows are good examples. Modelling of complete farming systems can also give valuable contributions in this area. The development of such feed evaluation systems should get a high priority because it is a prerequisite for a more sustainable animal production based on high input technology. An alternative option would be animal production on the basis of a low input technology, which only seems possible if the present animal industry in The Netherlands is reduced by at least half its present size.

Table 6.7. EFFECT OF pST ON TOTAL AND MARGINAL EFFICIENCY OF PROTEIN DEPOSITION IN GROWING PIGS[a]

Treatment	*CP*	*Intake*	*Deposition*	*Efficiency*	
	%	— g N/day —		*Total*	*Marginal*
Control[b]	16	369	149	40.3	
	18	417	159	38.1	21.0
	20	460	168	36.5	21.0
	18	414	159	38.4	
	20	461	164	35.6	
pST[c]	16	369	184	49.9	
	18	415	206	49.7	47.6
	20	457	225	49.2	45.0
	18	413	211	51.1	
	20	456	222	48.6	25.3

[a] Adapted from Van Weerden (1989)
[b] Energy intake 1100kJ of metabolizable energy per kg $W^{0.75}$
[c] Energy intake 1290kJ of metabolizable energy per kg $W^{0.75}$

Financial consequences

The aim for the year 2000 is reducing the input in animal feed of P by 40% and of N by 27% and reducing NH_3 loss by 70% compared to 1988. Solutions to reach this aim are considered technically possible, but will be extremely costly. Total costs are estimated at over 3 billion guilders. Compared with the investments already made this is an increase of about 2.5 billion guilders, the equivalent of one third of the total added value of animal production in The Netherlands (Wijnands *et al.*, 1991). This seems a tremendous challenge, but no other option may be feasible. About one quarter of the total extra costs or the equivalent of 560 million guilders are for animal feed. This will make the feed more expensive by about 4 guilders per 100 kg, an increase of 10 to 15%.

Legislation

From the previous section it becomes apparent that measures to reduce the release of undesired components from animal waste are economically unattractive. Enforced legislation to limit and reduce environmental pollution through farm animals seems therefore unavoidable. Regulations to limit emissions of gaseous and other pollutants by farm animals have already or soon will be introduced. Regulations to limit expansion of farm capacity has already been enforced. Storage of manure is also subject to severe restrictions. In addition the application of manure as a fertilizer will become limited.

In The Netherlands, legislation to reduce the application of excessive levels of phosphorus (P) through manure as a fertilizer was introduced some years ago, resulting in maximum application rates of manure. Present maximum rates are 200, 200 and 125 kg P/ha for land used for the production of grass, corn and other crops respectively. Before the end of the century, they will almost certainly be reduced further to 110, 75 and 70 kg/ha respectively. Surplus manure is being taxed and distribution (as fertilizer or as raw material for a manure refinery) is through a manure bank.

Regulations to reduce NH_3 emissions can be summarized as follows. Expansion of animal production enterprises is restricted in the vicinity of forests and nature reserves. Introducing a maximum stocking density is under consideration. Manure storage pits need to be covered in order to reduce emission of NH_3. Application of manure as a fertilizer is illegal between October 1 and February 1 and on snow covered land. On sandy soils application of manure as a fertilizer between June 15 and October 1 may only take place through soil injection.

Conclusions

Animal production systems require not only N deposited in the animal to be valued, but N excreted in waste needs to be taken into account as well.

Important gaseous pollutants originating from farm animals are fermentation gases, respiration gases, ammonia, nitrogen oxides and noxious odours. The first two gases are produced directly by farm animals, whereas the others result from microbial conversions during storage. Gaseous emissions can be considerably reduced through improved feeding management, improved ways of storage and proper ways of application of manure as a fertilizer.

In all species of farm animals excretion of N in waste can be minimised by reducing N input. Feeding management, like phase feeding, more frequent feeding, feeding totally mixed rations and matching rates of rumen degradation of proteins and carbohydrates are means for a further optimization of animal production and can reduce the loss of undesired end products to the environment. Utilisation of ingested N can be improved by balancing the protein to energy ratio, by balancing the amino acid ratio and by (bio-)technological treatment.

Because of interrelationships between energy and protein, requirements for them should not be considered separately. Factorial feed evaluation systems are inadequate for producing the response predictions which are needed to quantify the distribution between N deposition and N waste. More dynamic systems, taking into account the dynamics of the digestive and nutrient partitioning systems in farm animals should be developed.

Measures to reduce the excretion of undesired compounds to the environment are quite costly. Such measures are therefore not economically profitable and their introduction will have to be enforced by legislation.

References

Agricultural and Food Research Council (1991). *Nutrition Abstracts and Review, Series B: Livestock Feeds and Feeding*, **61**, 683–722

Agricultural Research Council (1981). *The Nutrient Requirements of Pigs.* Commonwealth Agricultural Bureaux, Farnham Royal, UK

Anonymous (1988a). In *Handboek voor de Rundveehouderij. (Handbook for dairy husbandry).* pp. 90. Ed. Pelser, L. Proefstation voor de Rundveehouderij, Schapenhouderij en Paardenhouderij (PR), Lelystad

Anonymous (1988b). In *Zorgen voor Morgen. Nationale Milieuverkenning 1985-2000. (Worry and Care for Tomorrow. National Environental Exploration 1985-2000).* pp. 84–155 Ed. Langeweg, F. Samson, Alphen a/d Rijn

Anonymous (1991). *Eiwitwaardering voor Herkauwers: Het DVE Systeem. (Protein Evaluation for Ruminants: The DVE System)*, Centraal Veevoeder Bureau nr. 7, Lelystad

Bauman, D.E. (1984). In *Herbivore Nutrition in the Subtropics and Tropics.* pp. 505–525, Eds. Gilchrist F.M. and Mackie, R.I., The Science Press, Craighall, RSA

Campbell, R.G. (1988). *Nutritional Research Reviews*, **1**, 233–253

Coleman, R.N., Feddes, J.J.R. and West, B.S. (1991). In *Proceedings Western Branch Meeting. Canadian Society of Animal Production*, Chilliwack, May 27-28. (Abstr.)

Coppoolse, J., Van Vuuren, A.M., Huisman, J., Janssen, W.M.M., Jongbloed, A., Lenis, N.P. and Simons, P.C.M. (1990). *IVVO Mededeling No. 3 (Update 1990).* Institute for Livestock Feeding and Nutrition (IVVO). Lelystad

Esteban Turzo, P., Gonzalez Fernandez and Garcia de Bustos, J.M. (1988). In *Volatile Emissions from Livestock Farming and Sewage Operations*, pp. 170–177. Ed. Nielsen, V.C., Voorburg, J.H. and L'Hermite, P. Elsevier Applied Science, London

Everts, H. and Dekker, R.A. (1991). Report 230. Institute for Livestock Feeding and Nutrition (IVVO-DLO), Lelystad

Fuller, M.F. (1991) In *Protein Metabolism and Nutrition, Volume 1.* pp. 116–126. Eds. Eggum, B.O., Boisen, S., Borsting, C., Danfaer, A. and Hvelplund, T. National Institute of Animal Science, Foulum

Fuller, M.F., McWilliam, R., Wang T.C., and Giles, L.R. (1989) *British Journal of Nutrition*, **62**, 255–267

Goossensen, F.R. and Meeuwissen, P.C. (1990). *Bijdrage van de Nederlandse land- en tuinbouw aan het broeikaseffect. (Contribution of Dutch agri- and horticulture to the greenhouse effect).* Informatie en Kenniscentrum Veehouderij, Ede

Huisman, J. (1990). *Antinutritional Effects of Legume Seeds in Piglets, Rats and Chickens.* Ph.D. Thesis, Agricultural University, Wageningen.

Ingram, D.L. and Dauncey, M.J. (1986). In *Control and Manipulation of Animal Growth,* pp. 5–20. Eds. Buttery, P.J., Haynes N.B. and Lindsay, D.B. Butterworths, London

Johnson, D.E., Hill, T.M., Carmean, B.R., Branine, M.E., Lodman, D.W. and Ward, G.M. (1991). In *Energy Metabolism of Farm Animals.* pp. 376–379. Eds. Wenk, C. and Boessinger, M. Institut für Nutztierwissenschaften, Zürich

Klarenbeek, J.V. and Bruins, M.A. (1988). In *Volatile Emissions from Livestock Farming and Sewage Operations.* pp. 73–84. Eds. Nielsen, V.C., Voorburg, J.H. and L'Hermite, P. Elsevier Applied Science, London

Krawielitzki, K., Zebrowska, T. Schadereit, R., Kowalczyk, J. Hennig, U., Wunsche, J and Hermann, U. (1990). *Archives Animal Nutrition,* **40**, 25–37

Lamming, G.E. (1986). In *Future Production and Productivity in Livestock Farming: Science Versus Politics.* pp. 127–148. Ed. DSA Elsevier, Amsterdam

Lenis, N.P. (1989). *Netherlands Journal of Agricultural Science,* **37**, 61–70

MacRae, J.C., Buttery, P.J. and Beever, D.E. (1988). In *Nutrition and Lactation in the Dairy Cow.* pp. 55–75. Ed. Garnsworthy, P.C. Butterworths, London

Madsen, J. (1985). *Acta Agriculturae Scandinavicae (Supplement),* **25**, 9–20

Moughan, P.J. and Verstegen, M.W.A. (1988). *Netherlands Journal of Agricultural Science,* **36**, 145–166

Muir, G.D. (1977). In *Hazards in the Chemical Laboratory.* Ed. Muir, G.D. The Chemical Society, London

Pain, B.F., Thomson, R.B., Rees, Y.J. and Skinner, J.H. (1990). *Journal of the Science of Food and Agriculture,* **50**, 141–153

Pfeiffer, A. and Henkel, H. (1991). In *Digestive Physiology in Pigs.* pp. 126–131. Eds. Verstegen, M.W.A., Huisman, J. and Den Hartog, L.A., PUDOC, Wageningen

Rainelli, P. (1989). In *Agricultural and Environmental Policies: Opportunities for Integration.* OECD, Paris

Schutte, J.B. (1989). In *Nutrition and Digestive Physiology in Monogastric Farm Animals.* pp. 75–88. Eds. Van Weerden E.J. and Huisman, J. PUDOC, Wageningen

Simon, O. (1989). In *Protein Metabolism in Farm Animals.* pp. 273–366. Eds. Bock, H-D. Eggum, B.O., Low, A.G., Simon, O. and Zebrowska T. VEB Deutscher Landwirtschaftsverlag, Berlin

Spoelstra, S.F. (1978). *Microbial Aspects of the Formation of Malodorous Compounds in Anaerobicaly Stored Piggery Waste.* PhD Thesis, Agricultural University, Wageningen.

Swanson, E.W., (1982). In *Protein Requirements for Cattle.* pp. 183–197. Ed. Owens F.N. Miscellaneous Paper 109, Oklahoma State University, Stillwater

Tamminga, S. (1991). *Journal of Dairy Science* (in press).

Tamminga, S. (1992). *Proceedings Internationl Conference on Farm Animals and the Environment.* CAB (in press)

Tamminga, S., Van Vuuren, A.M., Van der Koelen, C.J. Ketelaar, R.S. and Van der Togt, P.L. (1990a). *Netherlands Journal of Agricultural Research*, **38**, 513–526

Tamminga, S., Van der Poel, A.F.B., Savelkoul, F.H.M.G., Schutte, J.B. and Spoelstra, S.F. (1990b) In *Agricultural Biotechnology in Focus in the Netherlands.* pp. 155–164. Eds. Dekkers, J.J., Van der Plas H.C. and Vuijk, D.H. PUDOC, Wageningen

Tamminga, S. and Verstegen, M.W.A. (1991). In *Protein Metabolism and Nutrition, Volume 1.* pp. 23–36. Eds. Eggum, B.O., Boisen, S., Borsting, C., Danfaer, A. and Hvelplund, T. National Institute of Animal Science, Foulum

Valk, H., Klein Poelhuis, H.W. and Wentink, H.J. (1990). *Netherlands Journal of Agricultural Science*, **38**, 475–486

Van Bruchem, J. Bongers, L.J.G.M., Lammers-Wienhoven, S.C.W., Bangma, G.A. and Van Adrichem, P.W.M. (1989). *Livestock Production Science*, **23**, 317–327

Van der Meer, H.G. (1991) In *Mest en Milieu in 2000 (Manure and Environment in 2000).* pp. 15–24, Ed. Verkerk, H.A.C. Directie Landbouwkundig Onderzoek, Wageningen

Van der Poel, A.F.B. (1990). *Advances in Feed Technology*, **4**, 22–34

Van der Poel, A.F.B., Blonk, J., Huisman, J. and Den Hartog, L.A. (1991). *Livestock Production Science* (in press)

Van Es, A.J.H. (1980). In *Protein Deposition in Animals.* pp. 215–224. Eds. Buttery P.J. and Lindsay D.B. Butterworth, London

Van Straalen, W.M. and Tamminga, S. (1990). In *Feedstuff Evaluation.* pp. 55-72 Eds. Wiseman, J.and Cole, D.J.A. Butterworth, London

Van Vuuren, A.M., Tamminga, S. and Ketelaar, R.S. (1986). *Netherlands Journal of Agricultural Science*, **34**, 457–467

Van Weerden, E.J., (1989). In *Nutrition and Digestive Physiology in Monogastric Farm Animals.* pp. 89–101. Eds. Van Weerden, E.J. and Huisman, J. PUDOC, Wageningen

Verité, R. and Peyraud, J.-L. (1990). In *Ruminant Nutrition.* pp. 33-48. Ed. Jarrige, R. INRA, Paris

Vertregt, N. and Rutgers, B. (1988). In *Volatile Emissions from Livestock Farming and Sewage Operations.* pp. 85–91. Eds. Nielsen, V.C., Voorburg, J.H. and L'Hermite, P. Elsevier Applied Science, London

Wijnands, J.H.M., Oudendag, D.A., Luesink, H.H. and Van der Veen, M.Q. (1991). In *Mest en Milieu in 2000. (Manure and Environment in 2000)*. pp. 115–136. Ed. Verkerk, H.A.C., Directie Landbouwkundig Onderzoek, Wageningen

Wunsche, J., Hermann, U., Meinl, M., Hennig, U., Kreienbring, F. and Zwierz, P. (1987). *Archives of Animal Nutrition*, **37**, 745–764

Zeeman, G. (1991). *Mesophylic and Psychrophylic Digestion of Liquid Manure.* PhD Thesis, Agricultural University, Wageningen

7

REPORT ON THE EC COMMISSION ON THE USE OF GROWTH PROMOTERS AND OTHER ADDITIVES IN ANIMAL FEEDS - SOME PERSONAL OBSERVATIONS

K. LAWRENCE

Elanco Animal Health, Dextra Court, Chapel Hill, Basingstoke, Hants RG21 2SY, UK

Introduction

In 1989, the Division of Research in Agriculture of DGVI of the EC Commission sought tenders for a survey on the use of growth promoters in animal feeds. The contract to undertake the study was won by CEAS Consultants (Wye) Ltd, Wye College, University of London, the Agricultural Economics Department Wye College, University of London and the Institut für Physiologie, Physiologische Chemie und Ernährungs-physiologie, Ludwig-Maximilians-Universität, München. Collaborating with the primary contractors were, the Department of Applied Biochemistry and Food Science, University of Nottingham School of Agriculture, the Institut für Medizinische, Mikrobiologie Infektions und Seuchenlehre, Ludwig-Maximilians-Universität, München and Produce Studies Ltd.

The survey, entitled "The impact on animal husbandry in the European Community of the use of growth promoters in animal feeds", was initally limited to products listed in the Annexes to EC Directive 70/524 (see Table 7.1) and copper. However, the study expanded to include other classes of pharmaceuticals such as bST (bovine somatotrophin), pST (porcine somatotrophin) and β-agonists (e.g. clenbuterol).

The subjects included in the study and how the work was divided amongst the collaborators are listed in Table 7.2.

To enable accurate figures on the volume of growth promoter usage and the economic effects on production, husbandry systems were only examined in detail in a number of EC countries as set out in Table 7.3.

The final draft report was delivered to DGVI in February 1991 and the final report was submitted in July 1991. It has now been accepted by the EC Commission, but the contents still remain confidential. Since the report is not yet in the public domain, it is not possible to meet the initial objectives of this chapter. Consequently, rather than make superficial observations about what might appear in all the sections of the report, this brief chapter will concentrate on a personal perspective of just one aspect of the use of growth promoters, namely their effect on environmental pollution.

Table 7.1. GROWTH PROMOTERS LISTED IN THE ANNEXES TO EC DIRECTIVE 70/524

	Species covered in listing			
Chemical Name	*Pigs*	*Broiler Chickens*	*Cattle Beef*	*Cattle Dairy*
Avilamycin	+	+		
Avoparcin	+	+	+	+
Bambermycin	+	+	+	
Copper	+			
Monensin			+	
Salinomycin	+			
Spiramycin	+	+	+	
Tylosin	+			
Virginiamycin	+	+	+	
Zinc Bacitracin	+	+	+	

Environmental impact of the use of growth promoters

A summary of 59 European trials, involving 5551 pigs, showed that growth promoters increased the average daily gain by 21g/d and improved the feed conversion by -0.14 kg feed/kg gain (see Table 7.4). In addition, the use of growth promoters can mean a reduction of at least one percentage point in protein levels in pig rations (Yen and Veum, 1982; Gropp, 1989; Gropp and Birzer, 1990).

Such figures raise the important question of what is the potential environmental protection for the UK of an improvement of -0.14 kg feed/kg gain in FCR, for pigs grown from 10 to 85 kilogrammes body weight? For each pig there will be a reduction in total feed consumed of some 14 kg, with a decrease in protein consumption of 2 kg. Over 42 litres (9.2 gallons) less drinking water will be consumed and 43 litres (9.5 gallons) less slurry produced. Collectively these changes will lead to a reduction in nitrogen excretion of 0.3 kg and in phosphorus excretion of 0.1 kg in the faeces and urine.

Projecting these environmental benefits per pig, to a UK national kill of some 140 million pigs over the last 10 years, the likely benefits of the use of growth promoters have been:

Less feed consumed	1,470,000 t
Less protein consumed	206,316 t
Less water consumed	4 410 000 m^3
	(897,470,000 gallons)
Less slurry production	4,459,500 m^3
	(980,000,000 gallons)
Less nitrogen excreted	31,000 t
Less phosphorus excreted	10,300 t

These figures are certainly large, but just what do they mean in reality? Nearly one and a half million tonnes of pig feed saved is over 60,000 lorry journeys, each carrying 25 tonnes of feed, with their own effects on environmental pollution. The millions of gallons of water saved are nearly all from the mains - how much more acute the recent water shortages could have been, especially in Humberside and East Anglia, if growth promoters had been banned. The slurry saved over the last 10 years represents a pool one foot deep, covering an area of approximately 120 square miles. That slurry would also have had to have been transported and spread on to agricultural land, again having an indirect effect on the environment through burning up fossil fuels.

What are the overall effects within Western Europe, including pigs grown to 100 kg bodyweight? The calculated yearly savings are presented in Table 7.5. It is evident that the use of digestive enhancers produces a substantial saving in feed and water intakes, thereby reducing waste production and positively contributing to the protection of the environment.

The beneficial effects of growth promoters and BST on methane production (a greenhouse gas which also contributes to ozone depletion) from cattle must also be considered. Crutzen *et al.* (1986) identified cattle as the second largest anthropogenic methane source. The methane is formed as a consequence of the production of the volatile fatty acids, acetic and butyric acid. Leng (1992) estimated that the use of BST in the USA could reduce methane emissions by as much as 60% per gallon of milk.

With rumen active growth promoters, such as the ionophores (monensin sodium, Romensin® - Elanco Animal Health), model calculations by Bent *et al.*, (1992) estimate a reduction of up to 30 million cubic metres of methane per year from UK beef cattle.

The withdrawal of growth promoters approved for cattle would immediately increase methane emissions by 11-20%. This increase is double the target of 40-160 Gt reduction per year to halt the global rise.

Summary

With the accelerating population growth predicted up to and beyond the year 2000, the efficient use of the planet's resources must be our clear aim. As recently suggested by Dr. Helmut Schalz, the Permanent Secretary to the German Ministry of Food, Agriculture and Forestry, we must minimise the environmental pollution associated with animal production, especially of pigs and poultry. Two ways to achieve this end are (i) reduc-

Table 7.2. TOPICS COVERED AND COLLABORATORS INVOLVED IN THE EEC COMMISSION ON THE USE OF GROWTH PROMOTERS AND OTHER ADDITIVES IN ANIMAL FEED

	Topic	Collaborators
1.	Economic consequences for the economy of the EC by using growth promoters.	CEAS Consultants Wye College University of München
2.	Long term safety of growth promoters including environmental effects	University of München
3.	growth promoters and meat quality	University of Nottingham
4.	Consumer acceptance market research	Produce Studies Ltd.
5.	Socio-economic consequences for small farmers	CEAS Consultants Wye College
6.	Mode of Action	University of München University of Nottingham
7.	Implications for education and training for non-traditional growth promoters	CEAS Consultants University of München University of Nottingham

Table 7.3. DISTRIBUTION OF ANIMAL PRODUCTION SYSTEMS INCLUDED IN THE EEC COMMISSION SURVEY

Country	*Veal*	*Beef*	*Growing/ Fattening Pigs*	*Poultry (Broilers and Layers)*
UK		+	+	+
France	+	+	+	+
Italy	+	+	+	+
West Germany	+	+	+	+
Spain			+	+
Netherlands	+		+	+
Denmark			+	

Table 7.4. THE EFFECT OF DIGESTIVE ENHANCERS ON WEIGHT GAIN AND FEED CONVERSION IN PIGS (POOL OF 59 EUROPEAN TRIALS INVOLVING 5551 PIGS)

	Negative Control	*Growth Promoter*	*Benefit*
Average Daily Gain (g/d)	650	671	+21
Feed Conversion Ratio (kg feed/kg gain) (Weight range 28-98kg)	3.28	3.14	-0.14

Elanco (unpublished data)

Table 7.5. POTENTIAL YEARLY SAVINGS AND REDUCTIONS IN WASTE PRODUCTION FROM USING GROWTH PROMOTERS

	Due to improved FCR of -0.14kg feed/kg gain	*Due to a 1% Protein Saving Effect*
Possible Savings		
Feed consumption (t)	2,275,000	
Protein consumption (t)	292,500	533,000
Drinking water (m^3)	6,825,000	
Possible Waste Reduction		
Manure (m^3)	6,987,500	
Nitrogen (t)	48,750	84,500
Phosphate (t)	16,250	

Kroeger (1991)

ing the protein content in finishing feeds to meet demand and (ii) the use of licenced growth promoters. Indeed, it has been suggested that "Not to use growth promoters is an environmental crime" (Gropp and Birzer, 1990).

References

Bent, M., Birzer, D., and Gropp, J. (1992). (In press)

Crutzen, P.J., Aselmann, I., and Seiler, W. (1986). *Tellus*, **38B**, 271–284

Gropp, von J. (1989). *Ernaehrungsdienst*, **67**, 44–47

Gropp, von J. and Birzer, D. (1990). *Vet*, **6**, 90

Kroeger, H. (1991). Benefits of Pig Performance Enhancers for the Environment. *Proc. FEFAC Conference*, Paris, France

Leng, R.A. (1991). *Improving Ruminant Production and Reducing Methane Emissions from Ruminants by Strategic Supplementation.* Environmental Protection Agency, Washington

Yen, J.T., and Veum, T.L. (1982). *Journal of Animal Science*, **55**, 1099–1108

8

PREDICTING THE RESPONSE TO VARIATION IN DIET COMPOSITION

R.H. FAWCETT, M. WEBSTER[2], P.K. THORNTON, SHII-WEN ROAN and C.A. MORGAN

University of Edinburgh, Institute of Ecology and Resource Management, Kings Buildings, West Mains Rd, Edinburgh EH9 3JG.

[2] *Format International Ltd., Format House, Poole Road, Woking, Surrey GU21 1DY*

Introduction

This research grew out of a need to answer two simple questions:-

1. How accurately can the feed industry manufacture a compound feed to a customers specification?
2. What are the consequences of error and variability for the customer?

The stimulus came from the continuing activity of Shii-Wen Roan, Colin Morgan, Colin Whittemore, Gerry Emmans, John Oldham and others in modelling energy and protein utilization of farm animals. Since many people will apply such models to the design of feeding systems on farms it is necessary to point out that diet composition, which is a major instrumental variable in the models, is subject to a degree of uncontrollable variation confounding commercial monitoring information and any long term experiments.

The initial questions are directly relevant to quality standards and the limits to variation in Feedingstuffs Regulations. At present the regulations are under review. Such a review should question whether or not there is merit in reducing tolerances. The limits here are strongly influenced by the accuracy and precision attainable in current laboratory practice and calibration of instruments.

Manufacturers will be concerned with the practical steps which can be taken to reduce the variability of their products whilst feeders will be more concerned with the trade offs between increased specification, variance reduction, diet cost and improved animal performance.

A user oriented definition of 'quality' in a purchase is suitability for the purpose intended. Excesses or deficiencies in the nutritional value of feeds may have serious financial implications. Hence limits to variation must be recognized as an important aspect of quality control in feed manufacture.

This chapter follows through the integrated process from raw material delivery at a feed mill to on farm feeding. Animal growth models are used to estimate the benefits from some quality control practices.

During a two week period in January 1991 all bulk deliveries of raw materials to a feed mill were subject to Near Infra-Red (NIR) analysis and recorded. During the second week the continuous production log recording the manufacture of every batch of feed was subject to analysis of differences between formula weight and recorded weight for each ingredient. Samples taken from departing bulk delivery vehicles were subject to NIR analysis. The product samples were cross referenced to the formulation which generated them. Samples of raw materials and products were sent to other laboratories to check the NIR for crude protein and oil. Starch and sugar analyses were requested to complete the set for estimates of metabolizable energy via the EEC equation in the feedingstuffs regulations.

A statistical model is used to make theoretical predictions of the variability of the products from estimates of the variability of raw materials, weighing errors and analytical errors in raw materials and the compound feed.

Historical Perspective

There is not a lot of published work on the accuracy of feed manufacture or the variability of feed ingredients. In attempting to link both these factors together Burdett and Laws (1979) and Duncan (1989) commented that variation in the raw material crude protein content only accounts for twenty five or thirty percent of the variation in the composition of the product. Analytical errors were directly estimated by Burdett and Laws so the estimate for manufacturing variation is confounded with any residual there.

Clarke (1976) comments on raw material variation from a feed manufacturers perspective. In this presentation the variability of different sources is shown and the asymmetry of the distributions is evident. In general the distribution in reported contents is not normal. The fitting of a better description of an unknown distribution is a problem requiring large data sets. Better descriptions of variability are worth having because they would facilitate predictions of the consequences of fractionation before investing in the necessary equipment.

The assumption of normality may well be valid for analytical error but may not be appropriate as a description of variation in naturally occurring phenomena such as the chemical composition of raw materials. The assumption of normality in a distribution is often grasped because proof of the identity of a generating function for any small data set is a difficult task. Normally distributed protein content has been assumed in the literature.

One way of dealing with variability was to write down the mean by one half a standard deviation after Nott and Combs (1967). This is a pragmatic approach with desirable consequences but rather inefficient in achieving the objective. In feed formulation Chen (1973) used 'Quadratic Programming' for probability constraints taking into account raw material variation as the sole source of variation in the product; analytical errors were ignored. The work of Burdett and Laws (1979) appears to be the basis of the discussion in Alderman (1985) but is a statement of a result for crude protein variability under the special circumstances of the investigation. It is not a statement of what is regularly achieved in practice.

Analytical errors did not escape the attention of Miles and Quackenbush (1955) who demonstrated the technique of working in net standard deviations. Lehrman and Bei (1975) working with net standard deviations assessed the payoff to reduced variability in terms of plateaued animal response of the Fisher, Morris and Jennings (1973) type.

In all the published work implicit belief in the absolute accuracy of weighing systems is astonishing. A maximum tolerance of 0.25% (based on full scale deflection) is mentioned in passing by Burdett and Laws(1979).

Raw materials

Any strategy for control of variability commences with raw material sourcing, delivery and storage prior to use. There is no reason why raw materials should not be bought by analyzed content subject to agreed tolerance standards. The introduction of rapid NIR analysis of raw materials enables the feed mill manager to apply acceptance or rejection criteria at the mill gate. Raw materials from different sources can be stored in different bins and bulk raw materials can be fractionated on the basis of NIR analysis and each fraction treated as a different material. Such variance reduction techniques have a cost implicit in the extra hassle and storage capacity required or reduction in the range of alternative raw materials stored.

The feed mill in question obtained bulk supplies of HiPro soya from two sources and held only one high quality fish meal in stock. Bulk supplies of wheat were fractionated into HiPro wheat and LoPro wheat using NIR on delivery. This fractionation has some interesting consequences due to the correlation between chemical components.

Of the 132 wheats delivered 77 were classed as LoPro and 55 HiPro dividing at 11.2% CP. HiPro wheats have higher oil and fibre content. Differences in ash content are not so pronounced. This effect is clearly visible in Figure 8.1. The ramifications of variation in second, third and fourth nutrients associated with variation in the chosen nutrient for fractionation are not appreciated by the U.K. feed industry at large. One reason is lack of published information on variance covariance properties of raw materials.

The variance reduction achieved in wheat is shown in Table 8.1 below. A further consequence may be higher energy concentration in HiPro wheat.

Table 8.1. FRACTIONATION OF WHEAT AT 11.2 CP%

	132 loads all wheat		*77 loads LoPro wheat*		*55 loads HiPro wheat*	
	Mean	*s.d.*	*Mean*	*s.d.*	*Mean*	*s.d.*
CP%	11.1	1.05	10.3	0.51	12.2	0.55
EE%	1.1	0.46	0.8	0.35	1.4	0.42
CF%	0.8	0.52	0.5	0.26	1.3	0.43
TA%	4.5	0.22	4.4	0.20	4.6	0.24

The variance covariance properties or correlations between chemical components of raw materials are not published. MAFF(1990) UK Tables of nutritive value and chemical composition are singular in their approach and sparse in data for some materials (n<=4). What is required is a vector of means and a set of variance covariance matrices for all the

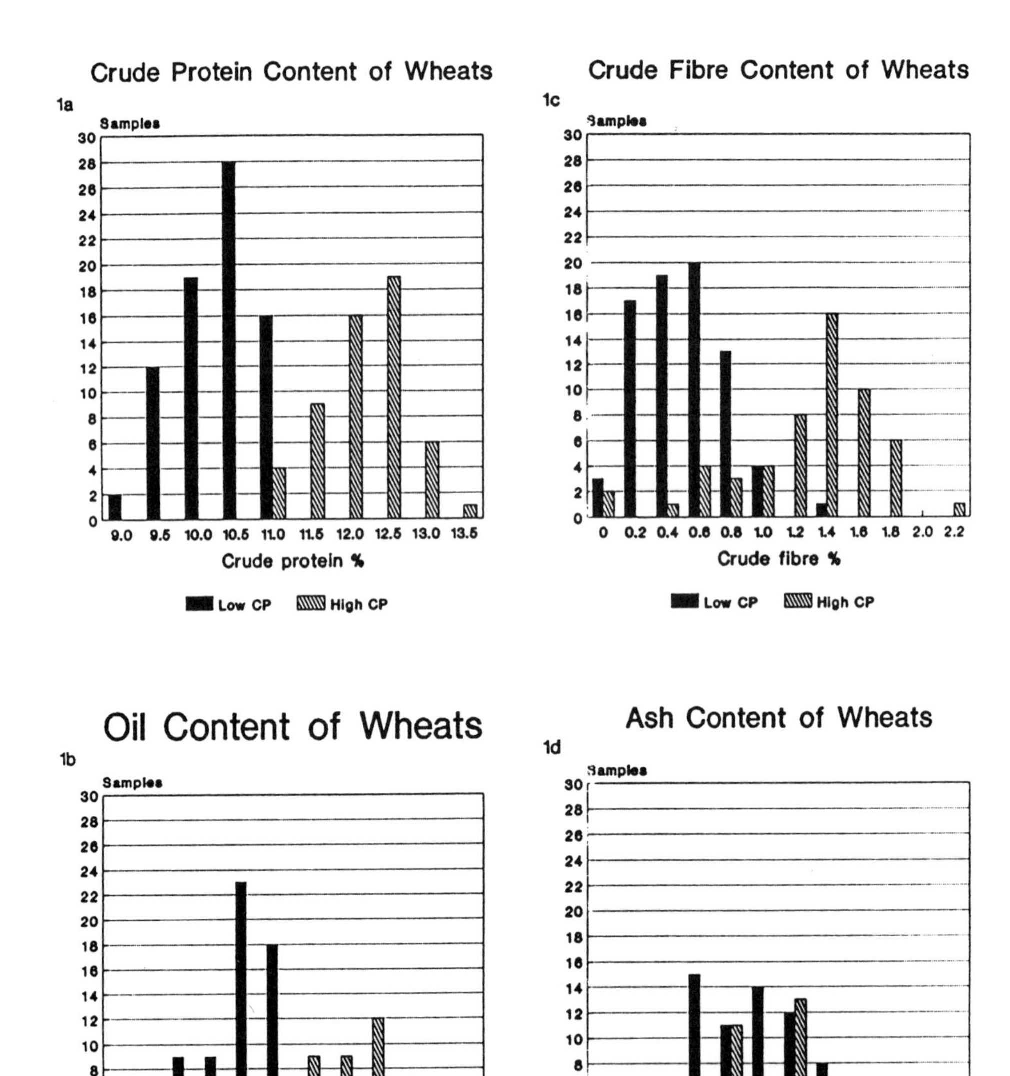

Figure 8.1. Effect of separating wheat supplies according to crude protein content or variation in other nutrients

components of the ME prediction equations published for all commonly available feed ingredients.

Air dry analysis is required by the feed mill and information would be better published in usable form rather than to suit the preferences of research workers. The MAFF (1990) table suggests that the crude protein content of U.K. wheats air dry will vary from 8.3% to 13.8% with a mean of 11% and standard deviation of 1.4% (n=45). Prediction of the consequences of fractionation requires at least the correlation matrix to accompany means and standard deviations of protein, oil, fibre and ash.

Table 8.2. CORRELATION IN THE COMPOSITION OF SOME 1990 WHEATS (n=132)

	CP%	EE%	CF%
EE%	0.548		
CF%	0.713	0.520	
TA%	0.053	0.435	0.559

The stability of the correlation matrix is of considerable interest for the prediction of the consequences of fractionation.

Formulation

Feed formulation is normally done by Linear Programming (LP). Account is taken of the volume of raw material stocks and demand for the products. Least cost optimization is performed for a given set of product specifications and tonnages required. If the linear programming matrix contains mean values for composition of raw materials the formula produced will only satisfy binding compositional constraints with a fifty percent probability. Nott and Combs (1967) recognized this. Probability constrained solutions can be obtained using quadratic programming or some iterative linear approximation to the implicit non linear constraints. Such probability constraining methods require detailed information on the variability of raw material sources. Utilizing the separation techniques outlined above, the linear programming matrix differs from the situation with unseparated materials in that it contains multiple raw materials of similar nature with slightly different compositions and reduced standard deviations.

Pricing fractionated raw materials can be problematic. Shadow pricing on the mean and standard deviation of reported contents offers a solution. Recombining fractionated products in equal proportions and thoroughly mixing will return the mean contents with reduced standard deviation.

The essence of the separation strategy is to accept a mean value for the contents of a storage bin and limit the variability of the raw material delivered into that bin. An alternative is to attempt to keep track of a moving average of bin contents. Unless homogenization of bin contents takes place this method offers no distinct advantages. A third possibility, adjusting formulation to NIR readings on raw material flow is a recipe for disaster, imposing a serious analytical and computational burden with frequent reformulation whenever significant deviation occurs from the previous matrix update.

Formulation errors

Formulation error can be said to exist when the raw material nutritional contents comprising a batch differ significantly from the values in the LP matrix. The advent of NIR does allow the manufacturer to set pre-defined limits to the magnitude of any formulation error.

Raw materials were coded according to the scheme at the feed mill. Means and standard deviations of reported nutritional contents were used to describe the raw materials. Formulae derived from the production log of the feed mill over the week long monitoring period were then dumped for testing alternative methods of calculating standard deviations and checking formulation errors on the 35 production formulae used in that week. Samples of the product from each formulation were analyzed by NIR. Examples of a Broiler feed and a Turkey feed are shown in Table 8.3.

Table 8.3. COMPONENTS OF VARIATION IN THE NUTRITIONAL COMPOSITION OF BROILER AND TURKEY FEEDS (SEE TEXT FOR DEFINITION OF TERMS)

				Broiler feed (n = 48)			
	CP%	*EE%*	*TA%*		*CP%*	*EE%*	*TA%*
EXP	21.49	9.40	5.32	ANA-EXP	-0.25	-0.05	+0.42
FOR	21.66	9.28	5.01				
RMs.d.	(0.32)	(0.19)	(0.18)	FOR-EXP	+0.17	-0.12	-0.31
ANANIR	21.24	9.35	5.74				
S.D.	(0.73)	(0.40)	(0.27)	ANA-FOR	-0.42	+0.07	+0.73
				Turkey feed (n=85)			
	CP%	*EE%*	*TA%*		*CP%*	*EE%*	*TA%*
EXP	24.28	7.83	6.62	ANA-EXP	-1.01	-0.80	+1.03
FOR	24.06	7.59	6.34				
RMs.d.	(0.29)	(0.17)	(0.11)	FOR-EXP	+0.22	-0.24	-0.28
ANANIR	23.27	6.99	7.65				
S.D.	(1.62)	(0.44)	(0.42)	ANA-FOR	-0.79	-0.60	+1.31

Across eleven feeds, seven Broiler and four Turkey, no large differences (FOR-EXP) have arisen between the FORMULAE multiplied by the raw material means and the EXPECTED result from the feed mill where the raw material file is updated monthly.

This suggests that variation in the mean composition of raw materials is small or tending to cancel out. The consistency of sign reveals slight bias in oil and total ash values i.e. all negative and less than 0.5%. There was no cause for concern over protein values, which exhibited 6 small absolute deviations less than 0.09%, and 3 positive and 2 negative absolute deviations between 0.10% and 0.25%.

There were however more significant systematic differences between the mean analysis of product samples and the expected analysis (ANA-EXP) or results of formula multiplication (ANA- FOR). Samples were drawn from each product and sent to two laboratories for checking.

The turkey feed returned a conundrum on crude protein (Table 8.4).

Table 8.4. CRUDE PROTEIN % (N=8)

	Mean	s.d.	
lab	23.49	0.833	Oct 91
NIR*	24.35	0.421	Jan 91
NIR**	21.36	0.522	May 91

The difference between NIR* and NIR** are the same machine at different times.

Highly significant differences between the means of samples and the EXPECTED outcome is a cause for concern but could be due to bias in the calibration of the NIR or change in operator.

Crude protein and oil should be limiting and may be expected to fall at the specified minimum. Ash may not be limiting but should be less than the specified maximum.

The large difference between the predicted standard deviations based on raw material variation alone (RMs.d.) using Chung and Pfost (1964) and the standard deviation (s.d.) from the analysis of samples of the product raises the question of what is the appropriate theoretical model for explaining variance in the final product.

Weighing errors

Milling and mixing is essentially a batch process. The formula consists of a set of weighing instructions for the control system to produce a batch of appropriate size. In feed mills dynamic accuracy is of the order of 0.5% of maximum scale deflection. The error stems from 'in flight weight'. There is a time lag between the cessation of feeding and the arrival of material on the scale. In flight weight depends upon the speed of the feeder, the flow characteristics of the material and the vertical distance between the feeder and the scale.

Flow characteristics of importance are particle size, density and any tendency towards coagulation. If the scale is full, the in flight column is less than when the scale is empty. The measurement of small quantities has proportionately greater errors. It is nonsensical to try and weigh small quantities on large scales hence in any feed mill there are a number of scales of varying size to minimize errors. Premixes are often added by hand and oil is often sprayed onto hot extruding pellets with some difficulty in obtaining an even distribution of the appropriate quantity.

Table 8.5. INGREDIENT S.D. SUBTRACTED FROM THE MEAN S.D.

	SCALE 1		
Formula	*Hi-wheat*	*Hyprosoya*	*Beans*
1	-	-	+
2	-	-	+
3	-	-	+
4	-	+	+
5	-	-	-
6	-	-	+
7	-	-	+
8		-	+
9		-	+
10		-	

			SCALE 2			
Formula	*Rape00*	*Canary*	*Sunflower*	*Peas*	*Meat&bone*	*Maize-germ*
1	-	-	-	+	+	+
2	-	-	-	+	+	+
3	-	-	+	+	+	+
4	-	+	-	+	+	+
5	-	-	-	+	+	+
6	-	-	-	+	+	+
7	-	-	-	+	+	+
8	-	-		+	+	
9				+		
10				+		

In any feed formulation algorithm it is recommended that minimum weighing quantities of five percent of maximum scale deflection be observed and a simple branch and bound procedure be used to adjust the formula. Rounding simply creates error.

Analysis of the production log of ten formulations with production runs of between twelve and thirty six batches weighing between 3.3 and 3.8 tonnes, comprising twelve or more ingredients, provides a sample of weighing errors from the mill. The computer log records the target weight set by the formula and the weight recorded by the digitised weight scale. The recorded difference between formula and scale weight of each ingredient is the raw data analyzed and is in turn subject to random errors unknown.

In the mill five scales were used. Some scales may be used to weigh a range of different ingredients but any ingredient is always weighed on the same scale. The mean difference between formula and delivery is a measure of the 'bias' of the scale which is a composite of a scale effect and an ingredient effect. Having several ingredients weighed over the same scale facilitated the separation out of ingredient effects. This is highlighted when the standard deviation of differences for individual ingredients is subtracted from the mean of the standard deviations. The signs of such differences are displayed in Table 8.5. For scale 1, the consistency of sign seems to indicate that the particle size and density of field beans produces an effect on column density markedly greater than that from wheat or soya. This is then reflected in the standard deviation in the batch weights of field beans in each formula. A similar pattern emerges with respect to scale 2.

The maximum recorded error standard deviation (Table 8.6) was 6.02kg on scale 1 compared with 3.25kg on scale 5 which is a reflection of the different materials weighed on 1 compared with 5. The wide range of materials weighed on 2 also explains the maximum error standard deviation of 4.36kg.

This suggests that each raw material should have its own individual weighing error in the computation of the theoretical variation in composition. The weighing error can thus reflect the size of the scale as well as the flow characteristics of the material. The particle size being important in the case of peas. No obvious explanation can be advanced for the high standard deviations observed in flour on scale 2.

Table 8.6. WEIGHING ERRORS (kg) IN RAW MATERIALS

Raw material	*(Scale)*	*Observed*		*dynamic*
		Mean s.d.	*Max s.d.*	*accuracy*
HiPro wheat	(1)	1.29	2.46	5.0
LoPro wheat	(5)	2.31	3.25	5.0
flour	(2)	1.51	4.36	2.5
HiPro soya	(1)	2.02	5.02	5.0
sunflower	(2)	1.15	1.93	2.5
rape-00	(2)	1.02	1.17	2.5
peas	(2)	2.27	4.46	2.5
beans	(1)	5.08	6.02	5.0
fishmeal	(2)	0.94	1.59	2.5
meat & bone	(2)	1.95	2.50	2.5
ppm	(2)	1.15	1.34	2.5
oats	(2)	1.08	1.50	2.5
maize-germ	(2)	1.70	1.97	2.5

Although oil and premixes were not weighed by the system it was possible to add in the constant balancing item to check the total weight in an identical manner to the computer control system which assumes no errors in the components added by the weigh scale coded zero.

The aggregate error standard deviation corresponds with the square root of the sum of error variances. The aggregate standard deviation will therefore increase with the number of weighings performed. The more ingredients the greater the weighing error.

In the absence of any static check weighing, the pooled standard deviation for each raw material is the best estimate that can be made for the weighing error attaching to individual feeds. It would be prudent to add something for the unknown error variance based on a static accuracy of 0.1% of full scale deflection. Alternatively the maximum observed standard deviation has validity in that it has been observed in a production run.

Mixing error is not considered since 'modern horizontal mixers have a minimum specification of dispersal to 1 part in 100,000 based on particles of like size and density' to quote Burdett and Laws (1979).

Accuracy and Precision in chemical analysis

Central to the whole purpose is accuracy and precision in chemical analysis. Precision relates to the clustering of results from repeated analysis on fractions of the same homogenized sample around a local mean and can be expressed as a standard deviation. Accuracy relates to the deviations of a local mean from a true mean or target. Accuracy can only be considered if you have valid alternative methods of measurement. The best that can be done is the standardization of methods in laboratory practice. Accuracy is unknown and probably does not matter as long as everyone uses the same protocol in analysis.

With the introduction of reference laboratories throughout Europe the characteristics of the analysis of samples delivered to different laboratories can be established. Regular ring testing procedures have been instituted to maintain the precision of reference laboratories. Independent research workers do not know how to interpret the results of ISO 5725 the protocol to determine repeatability and reproducibility for a standard test method by inter laboratory tests. There is a moral dilemma in routine analytical chemistry brought to our attention by Hamaker (1986) who demonstrates that workers are inclined to censor replicate test results so as to bring about convergence in test data. As a direct consequence the within lab variation may be understated by as much as fifty percent.

The weakness of open duplicates in the protocol of ISO 5725 is confirmed by Zaalberg (1989) who using hidden duplicates has shown that within lab variation was greater than the reported analytical variation derived from open duplicates by a factor of 1.7. Zaalberg (1989) goes on to show that ISO 5725 gives good enough results for between lab variation.

Even if the protocol were amended to provide hidden duplicates the method is defective in having zero degrees of freedom in the estimate of within laboratory variation. The Dutch protocol NEN 6303 uses twice the number of samples and produces a proper estimate of within laboratory variation. A comparison of the two methods reveals that ISO 5725 can overestimate within laboratory variation by a factor of 1.5 at high concentrations of the analyte. A proper estimate of within lab variation is what is required to establish the true variability of raw materials and compounds so there would be some logic in UKASTA adopting NEN 6303 protocol.

The advent of NIR spectroscopy provides for cost effective analysis on site. The speed of analysis makes it an ideal tool for process control. The nature of the technique is such that the machine requires calibration. One cannot expect the accuracy of NIR to be better than the laboratory technique used for calibration. However the repeatability standard deviation with NIR may well be less than the repeatability standard deviation within a laboratory using the calibrating methods. This poses a problem as to whether or not NIR results are qualitatively different from those produced by laboratory methods.

From the point of view of the user of results from an analytical service, laboratory guidance should be given as to the magnitude of the error variance which should be subtracted from the variance in the analysis of reported contents in order to establish the true variability of the product. Failure to work with true variability in any aggregation process leads to serious compounding of errors.

Alderman (1985) stated that existing usage tolerances are twice the standard deviation of the measurement under consideration hence analytical variances have been assigned the values shown in Table 8.7 for compound pig and poultry feeds.

Table 8.7. ANALYTICAL VARIANCES FOR COMPOUND FEEDS

crude protein%	0.2025	ash%	0.0625
ether extract%	0.0400	starch%	0.3906
crude fibre%	0.0506	sugar%	0.0625

Miles and Quackenbush (1959) indicated that analytical variation on straights is much less than on compounds and suggest a value of one half, in standard deviation units. For lack of suitable contemporary evidence a compromise on one half in variance units has been used in simulation work to date. Bailey and Henderson (1990) have published tables based on the percentage analyte concentration using Horwitz (1982). Reproducibility standard deviations are appropriate to regulatory work and repeatability standard deviations for netting out analytical variation from raw material variability.

The arguments for errors are probably best expressed as coefficients of variation to take account of scale effects in the wide range of raw material composition.

Previous work has taken an additive approach to the analysis of variance as if raw material variation was independent of the weighing system. This simplification would be valid if weighing errors were very small. However it has been demonstrated that weighing errors in the context of dynamic accuracy and variable raw material characteristics can not be ignored. More significant is the fact that the appropriate statistical model is multiplicative.

Statistical model of product variation

Weighing error and raw material variation net of any analytical variation, interact to produce variation in the composition of the product. In this situation raw material variation is amplified by the weighing errors to such an extent that reduction of raw material variation will play a significant role in reducing variation in the product.

Let there be n ingredients and the weight of the ith ingredient be Wi. Let there be k nutrients specified. Let the analysis of the ith ingredient for the jth nutrient be Aij.

Then the proportionate composition of the jth nutrient Pj can be defined as

$$Pj = \Sigma(WiAij)/\Sigma(Wi) \tag{8.1}$$

The equation 8.1 can be simplified to Z = X/Y by substituting Z for Pj, X for (Wi Aij) and Y for (Wi). The random variable X is the sum of products of random variables Wi and Aij.

The expectation and variance of products is discussed in Goodman (1960) and expanded upon in Bohrnstedt and Goldberger (1969). The random variable Y is in the sum of random variables Wi. The expectation and variance of the random variable Z is the result of the division of two random variables X and Y. There is little published information on the quotient result.

Let E[.] stand for the expectations operator, V[.] for the variance operator and COV[.] for covariance. Approximate formulae for expectation and variance can be written as

$$E[Z] \approx E[X/Y] \approx E[X] - COV[ZY]/E[Y] \tag{8.2}$$

$$V[Z] \approx E^2[Y]V[X] + E^2[X]V[Y] - 4COV^2[XY]/(E^2[Y])^2 \tag{8.3}$$

Simulation shows that COV[XY] tends to be small and the approximations are good enough to eliminate the need for a simulation within an algorithm, thus saving on computational burden. The variance of Z is considerably greater than the simple reduction formula applied when the weights are treated as constants. Previous writers on this subject have ignored weighing errors as a component of variance and some have ignored analytical variation.

The interaction of weighing error with net raw material variation will account for a significantly greater proportion of the variability of the reported contents of the product. Adding back the analytical variation in the product plus an estimate of the random residual provides a method for predicting the standard deviation in the manufactured product.

Cross contamination and partial separation before pelleting will contribute another source of variability which can not be parted from any random residual without properly designed statistical investigation. Theoretically the random fluctuations in the composition of formula ingredients should tend to compensate. Increased compensating variation can be purchased by either restricting maximum inclusion or using mixed integer programming to enforce inclusion of more than one ingredient from particular groups of feeds e.g. cereals or high protein meals.

The power of increasing numbers of ingredients to reduce variation is strictly limited. More ingredients included in a mix implies lesser weight of individual ingredients in the mix and greater significance of weighing errors.

Increasing the batch size of the mix reduces the effect of weighing errors but there are limitations in this direction as there will be tendency for the efficiency of mixing to decrease with the size of the mix unless longer time is allowed for effective mixing. If the required quantity rises above the capacity of the scale so that it is weighed in two drops the weighing error will be incurred twice. The batch size effect is reported in Fawcett and Webster (1991).

Nutritional value of compound feeds

For the purposes of describing the nutritional value of a compound feed the principle components are energy and protein content. With digestible energy or apparent metabolizable energy and crude protein, techniques of analysis are sufficiently precise to provide meaningful estimates of the parameters of a bivariate probability distribution. A nested set of ellipses can be identified as iso-probability contours in any bivariate diet space. The expected variance-covariance relationship between energy and protein can be established by simulating the multivariate properties of the compound feed and multiplying into the estimating equation for the energy content of compound feeds.

An example is shown in Figure 8.2. The covariance is responsible for the slope of the axes of the ellipse. With independence the covariance is zero and the axes of the ellipse are parallel to the axes of the diet space.

In vivo estimation of the energy value is both expensive and time consuming. The procedure requires a representative group of animals to be fed for several days. Food input, urine and faeces output are weighed and samples burned in a bomb calorimeter to determine gross energy content.

Prediction of energy value from regression equations based on chemical analysis of feed alone is cost effective for description, regulation and monitoring the quality of compound feed. The benefits of speed and cost reduction outweigh the negligible loss of precision.

Suitable regression equations have been agreed for the prediction of digestible energy (DE) for pigs and apparent metabolizable energy (AME) for poultry (Equation 8.5). The chemical analyses required are crude protein (g/100g), ether extract (g/100g), crude fibre (g/100g) for pig DE. Additional information on starch (g/100g) and total sugar (g/100g) is required for prediction of poultry AME. The unit (g/100g) is interpreted as a percentage (%) by weight as fed air dry in the equations.

$$DE(MJ/kg) = 0.1875\%CP + 0.3281\%EE - 0.1552\%CF + 0.1698\%NFE \qquad (8.4)$$

(cv%=3.8) (Henry *et al.* 1988).

$$AME(MJ/kg) = 0.1551\%CP + 0.3431\%EE + 0.1669\%ST + 0.1301\%TS \qquad (8.5)$$

(S=0.51 MJ/kg) (Alderman, 1985).

Cooke (1987) makes an impassioned plea for a limit to variability of ± 1MJ/kg when the EEC directive was asking for ± 0.4MJ/kg. Such is the power of UKASTA and FEFAC that the limits of variations in the poultry equation were set at +/- 0.7 MJ/kg (Statutory Instruments 1988 no 396 schedule 3 part E p36). Summation of the energy equivalent of four independent errors in chemical analysis at the 95% limit produced a worst case of +/- 0.81 MJ/kg. Exactly what the relevance of this procedure is in establishing the limits to variation requires an explanation by a statistician.

Bailey and Henderson (1990), using the same approach, suggested that the confidence interval attributable to analytical variation in the four determinations in the

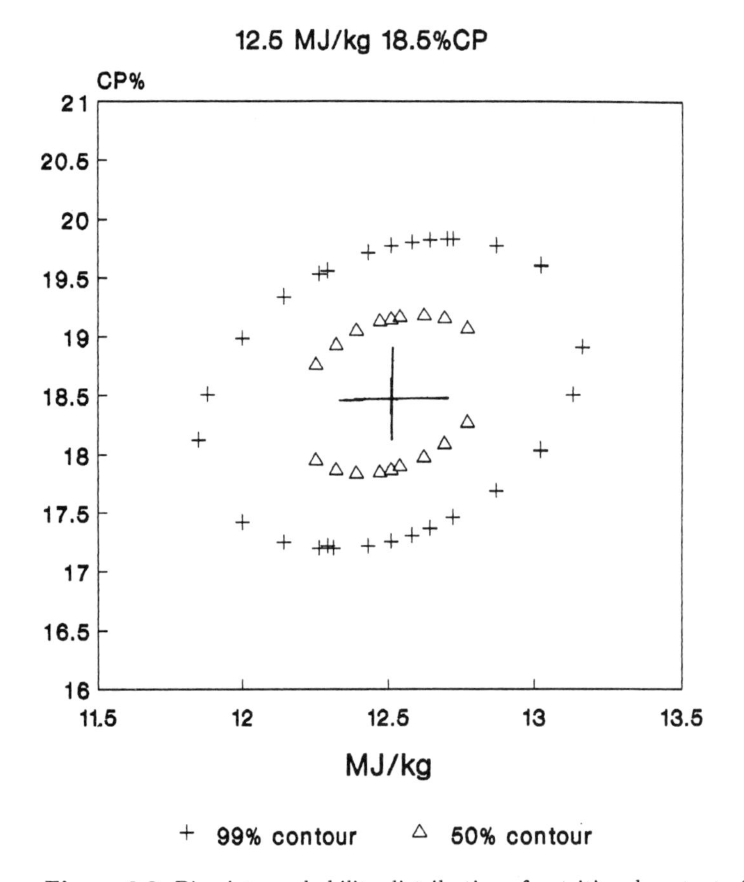

Figure 8.2. Bivariate probability distribution of nutritional content of compound feeds

poultry equation is of the order of +/- 0.4 MJ/kg, using relative standard deviation values from Horwitz (1982), for a diet containing 18.4% protein, 6.3% oil, 46% starch and 4% sugar and with an AME of 13.22 MJ/kg. The implication is that the performance of ADAS and UKASTA laboratories does not meet the standard expected by Horwitz (1982). What incentive is there for improved performance?

There is a dearth of published analyses of starch and sugar in raw materials, particularly relating to variability and the variance-covariance properties of the elements of the regression equations. The national data base is devoid of such information.

Fisher and McNab (1987) have shown that digestibility of oil varies with age in poultry. Yutste *et al.* (1991) have shown starches to vary in digestibility from 99% in wheat and cassava down to 72% in beans and 44% in potatoes. Whilst it is possible to formulate directly to a predicted DE or AME using the coefficients of the regression equations above, ingredient- specific digestibilities, with gross energy values, may produce better predictions for diets formulated out with the range of ingredient composition used in deriving the regression equations.

There are valid reasons why the imputed energy value of a raw material, obtained using the coefficients of the regression equations, may differ from the experimental value derived for a single raw material. If the normal method of consumption is in compound feed, imputing a raw material value from the shadow prices of chemical composition is logically sound. If the feed industry is to be regulated by prediction equations then compounders would be well advised to use them in setting energy constraints.

Predicting the response to variance reduction

The benefits of adopting a variance reduction technique should be measurable in terms of animal performance. The variation in feed composition due to natural variation in raw material composition is much smaller than that used to demonstrate a nutritional response experimentally. Massive research facilities are required to demonstrate the effects of variance reduction as reported in Duncan (1988). However, computer models are a suitable base from which to study the likely effect of small changes in the 'environmental' variables in a livestock production system.

With deterministic prediction models such as the 'Edinburgh Model Pig' or the 'Poultry Growth Model', the uncontrollable effects and errors of observation are effectively removed from the system under scrutiny and the pure response revealed. These models have been used to provide estimates of the potential benefits from reducing variation.

Given the operational characteristics of commercial fattening units and current price information it is possible to generate three dimensional financial response surfaces from factorial combinations of energy and protein consumed in a prescribed dietary regime. There are in fact no limits to the number of dimensions - three is the limit to what we can effectively manipulate visually. The variable price response surface and contours for Roan's (1991) pig model are shown in Figure 8.3a and 8.3b. Contour maps can be readily used to locate the profit-maximizing energy and protein combinations.

The feeder will receive a sequence of deliveries of the specified diet which exhibits between delivery variation at contracted price per tonne. Therefore a constant price response surface is required for exploring the consequences of variability in the specified diet. Given the cost of delivery of the profit-maximizing diet, it is possible to generate the response surface with constant feed cost. A constant cost surface for the dietary regime (Figure 8.3c) with contours (Figure 8.3d) simply reflects physical productivity and the effects of variation in energy and protein content on the grading of pigs.

Natural variation in the composition of the diet has, in one sense, zero cost since it does not cost anything to put it there. On the other hand there is a significant cost associated with reducing natural variation.

The 'expected' profit from feeding a target specification, of known variability, is obtained by integrating the bivariate probability distribution associated with that specification over the constant cost response surface.

The value of variance reduction can be estimated by modifying the parameters of the bivariate probability distribution and integrating over the constant price surface again. The difference in expected value is an estimate of the benefit deriving from any change in the parameters of the bivariate probability distribution. Pure variance reduction can be envisaged as the contraction of the probability contours towards the centre of the ellipse.

The shape of the response surface reflects diminishing returns so that the financial consequences of any negative deviations are not balanced by the financial consequences of

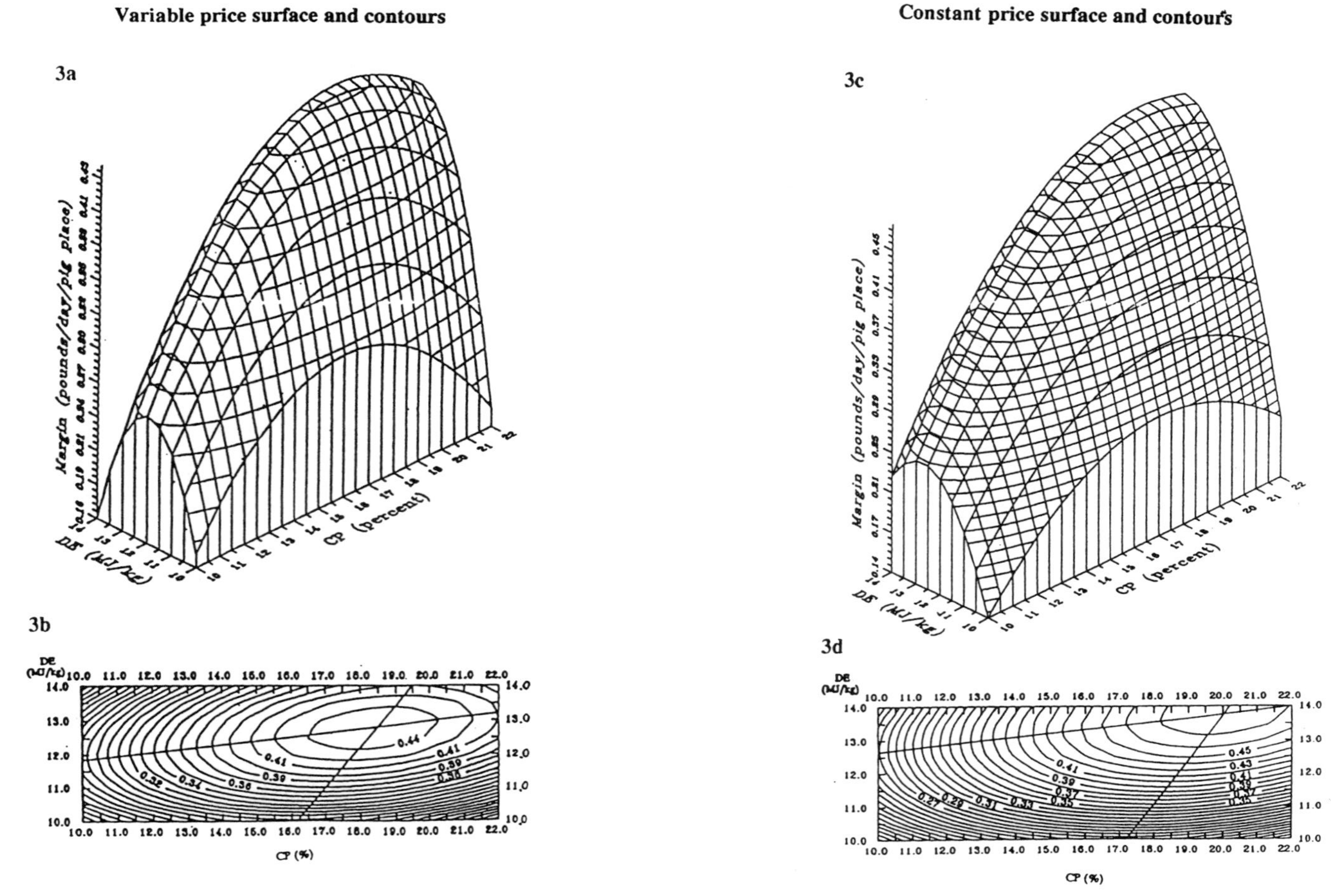

Figure 8.3: Response surface and contour maps from Shii-Wen Roan's pig model with variable or constant prices

positive deviations. The expected value is less than the value at the target specification. In this situation variance reduction increases expected profit.

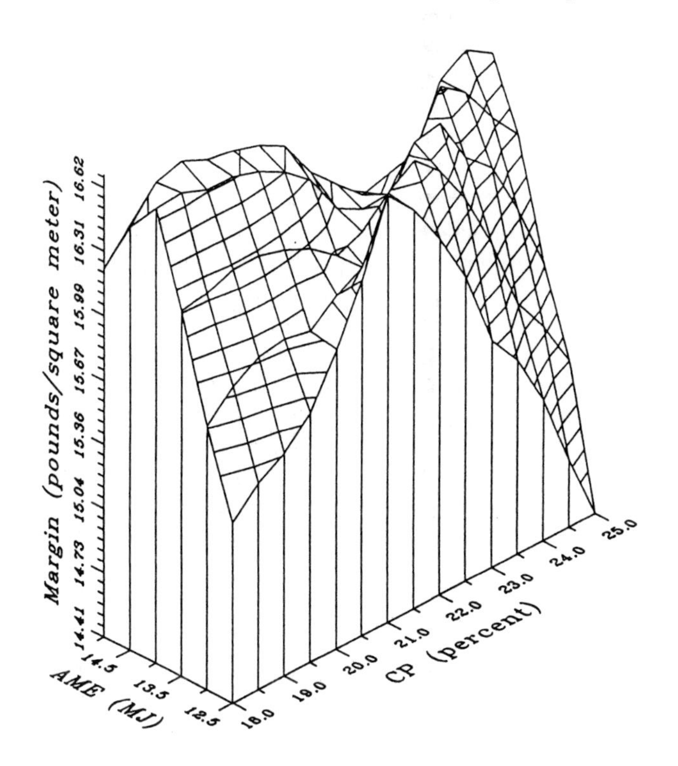

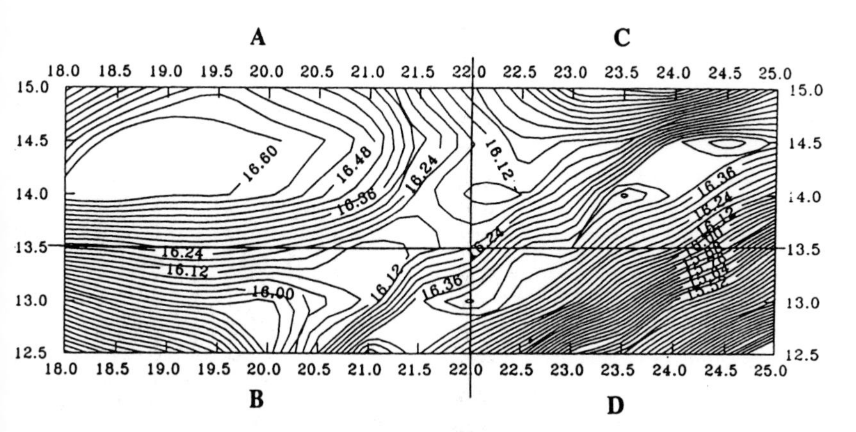

Figure 8.4. Variable price response surface (actual values) with contour map (from poultry model of G. C. Emmans)

The variable price response surface is not necessarily unimodal or smooth. If there are frequent changes in the basis of the LP solution over the response surface grid, this can lead to discontinuities in price changes. In the poultry growth model lysine is chosen as the limiting amino acid and fixed as a percentage of the crude protein. Marked discontinuities can be seen in the variable price response surface (Figure 8.4). The contour map can be divided into sections isolating the profit maximising peak from the other features of the irregular solid. Multivariate statistical forms can be moulded to fit the isolated surface and remove random variation. The fitted response surface (Figures 8.5a and 8.5b) can then be transformed to the constant price surface (Figures 8.5c and 8.5d) for evaluation of variance reduction within the section boundaries.

Preliminary results with pigs makes variance reduction worth up to £3.40 per tonne of feed consumed. The value from poultry has an upper limit of £1.56 per tonne. This major difference is because the poultry market does not have the penal grading system for fatness currently in force for pigs.

The cost effective way of exceeding target specifications is by a combination of increased mean and reduced variation, probably by bringing variance-reduced materials into the formula. There will be a market opportunity for bulk suppliers of raw materials to fractionate and sell variance- reduced material to the feed compounder.

Integrated feeding organizations will be looking towards a definition of the optimal level of variability in the diet since attempts to reduce the variability are bound to run into rapidly diminishing returns as a residual amount of variation is outwith the control of the manufacturer.

The context of on farm feeding

The bag is no longer the unit from which a sample is drawn. The contents of different bags could be from different batches weighed out in the mill. Between-bag variation could be of a magnitude approaching between-batch variation. Aggregation of samples from several bags ensured a single sample more representative of the load delivered. Within-delivery variation has not been a subject of reporting in feedingstuffs regulations but is relevant to the results of short duration feeding trials.

Bulk delivery and bulk storage is now the norm for 'on farm feeding'. Assuming that feed is made in three-tonne batches, a bulk delivery will consist of feed from at least seven batches. If we consider the progeny of a one-hundred sow unit the grower/finisher diet would be delivered in twenty-ton lots every fourteen days. At the feed mill the milling and mixing procedure will tend to homogenize the contents of each three- tonne batch. Drawing from several batches will put a within-load variation into the contents of the bulk delivery. What is fed on any day is a random drawing from this within-bulk variation. The parent distribution reflecting between-delivery variation will have new parameters generated every fourteen days.

Sampling from bulk delivery vehicles is done by taking three cupfuls from the flow of feed during loading. Observed variation from bulk delivery vehicles will be less than the between batch variation generated in the Mill because it will be the mean composition from three different batches. Whilst the product specification may remain constant, the formulae will not remain constant for more than a few days. Thus the variation between delivery at fourteen-day intervals in the feeding programme will have a between-formulae component.

The simulation of 'on farm feeding' requires a two stage sampling procedure. Within-delivery variation is experienced on a daily basis and between-delivery variation is

Isolated variable price response surface with contours

5a

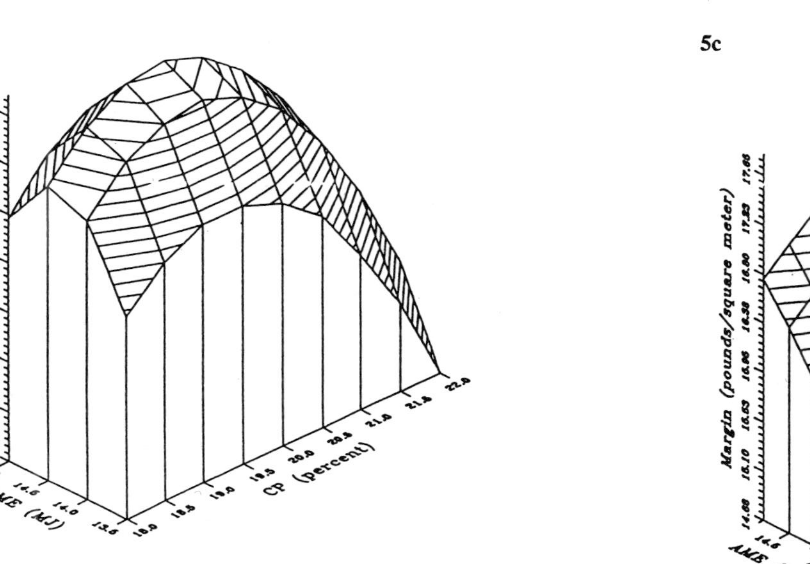

5b

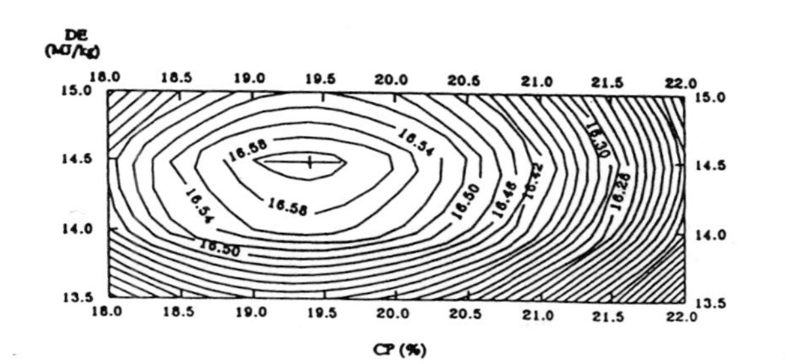

Isolated constant price response surface with contours

5c

Margin (pounds/square meter)

AME (MJ/kg)

CP (percent)

5d

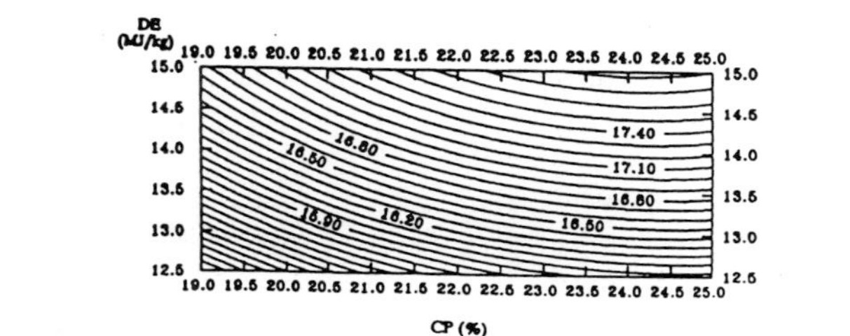

Figure 8.5: Response surface and contour maps from Emmans poultry model with variable or constant prices

experienced on a fortnightly basis. The results of such simulation does not lend itself to the generation of smooth response surfaces. However the cumulative probability distributions of financial outcomes can be generated and subjected to stochastic dominance analysis.

The question to be resolved is whether or not the variation in outcome is damped or exaggerated by multiple deliveries during the growing period. Stochastic versions of the models are required for such simulation studies with astronomic numbers of runs. Parallel processing on super computers now makes this a feasible proposition.

Better description of nutritional value

Crude protein is not a satisfactory description of the nutritional value of the nitrogenous component of the diet. To be consistent with the energy component, digestibility is the first modification which can be made. Digestibility between species is distinctly different. Digestibility increases with age within species. Digestibility appears to be site-specific, whether it be Edinburgh v Nottingham or Ileal v faecal.

Historically, the approach taken in research on responses to amino acids has been to first induce a deficiency and then demonstrate a response to a single limiting amino acid. However, increasing the concentration of a limiting amino acid simply moves the problem on to the next limiting amino acid or interactions, as demonstrated by D'Mello and Lewis (1970). There are also possible consequences of serious imbalance if too much of an amino acid is added or the supplement is not effectively dispersed.

For amino acid supplementation, the first step is to define the dietary intake of ideal protein for the particular stage of growth, pregnancy or lactation. What is ideal today may not be ideal tomorrow as far as the animal is concerned. Batterham (1980) has shown preferential absorption of synthetic amino acids which raises the question of whether or not synthetic amino acids can be used effectively in supplementation.

The amino acid intake requirement in the 1990's is expressed in ileal-digestible terms. The precision which a compounder can achieve in formulating to an ileal- digestible requirement is a function of analytical error and the natural variability of raw material sources. Such information is not readily available and may cost too much to acquire on a routine basis.

A multivariate statistical estimate of the means and variance- covariance properties of ileal digestibility of likely limiting amino acids for particular raw material sources can be made. These values can then be multiplied onto the mean and variance-covariance properties of amino acid concentration. Covariance between digestibility and concentration can not be ruled out. A probability density function can be derived for the ileal-digestible ideal protein. The crucial question is whether or not it is more profitable to feed on the basis of ileal-digestible amino acids rather than crude protein.

The limits to effective nutritional practice are not to be found in the lack of sophistication of biological models but in the techniques of chemical analysis, where cost and lack of precision may limit the profitable extent of application of nutritional knowledge.

Conclusion

Animal performance is influenced by unintentional variation in feed composition. Weighing errors and raw material variation interact to create significant variation in compound feeds. Financial benefits from enhanced animal performance may be obtained by reducing variability in feed composition. There may be known nutritional responses which cannot be effectively incorporated into feed formulation because cost of information and lack of analytical precision prevent effective quality control of some nutrients in commercial compound feed.

References

Alderman, G. (1985). In *Recent Advances in Animal Nutrition*, pp. 3–52. Eds. W. Haresign and D.J.A. Cole. Butterworths, London

Bailey, S. and Henderson, K. (1990). In *Feedstuff Evaluation* pp. 353–363. Eds. J. Wiseman and D.J.A. Cole. Butterworths, London

Batterham, E.S. (1980). In *Recent Advances in Animal Nutrition*, pp 11–22. Eds. W. Haresign and D. Lewis. Butterworths, London

Bohrnstedt, G.W. and Goldberger, A.S. (1969). *Journal of the American Statistical Association*, **64**, 1439–1442

Burdett B.M. and Laws B.M. (1979) In *Food Intake Regulation in Poultry*, pp. 405–429. Eds. K.N. Boorman and B.M. Freeman. British Poultry Science Ltd., Edinburgh

Cooke, B.C. (1987). In *Recent Advances in Animal Nutrition*, pp 19–26. Eds. W. Haresign and D.J.A. Cole. Butterworths, London

Chen, J.T. (1973). *American Journal of Agricultural Economics*, **55**, 175–183

Clarke, H.E. (1976). In *Proceedings of International Symposium on Computer Use in Feed Formulation*, pp. 14–19. National Renderers Association, Brussels

D'Mello, J.P.F. and Lewis, D. (1970). *British Poultry Science*, **11**, 367–385

Duncan, M.S. (1988). In *Recent Advances in Animal Nutrition*, pp. 3–11 Eds. W. Haresign and D.J.A. Cole. Butterworths, London

Fawcett, R.H. and Webster, C.M. (1991). *Feed Compounder*, **11** (10), 20–22

Fisher, C., Morris, T.R. and Jennings, R.C. (1973). *British Poultry Science*, **14**, 469–484

Fisher, C., and McNab, J.M. (1987) In *Recent Advances in Animal Nutrition*, pp. 3–18. Eds. W. Haresign and D.J.A. Cole. Butterworths, London

Goodman, L.A. (1960). *Journal of the American Statistical Association*, **55**, 708–713

Hamaker, H.C. (1986). *Journal of the Association of Official Analytical Chemists*, **69**, 417–482

Henry, Y., Vogt, H. and Zoiopoulos (1988). *Livestock Production Science*, **19**, 299–354

Horwitz, W. (1982). *Analytical Chemistry*, **54**, 67A–76A

Lerman, P.M. and Bie, S.W. (1975). *Journal of Agricultural Science*, **84**, 459–468

MAFF (1990). *UK Tables of Nutritive Value and Chemical Composition.* Rowett Research Services Ltd, Aberdeen

Miles, S.R. and Quackenbush, F.W. (1955). *Journal of the Association of Official Agricultural Chemists*, **38**, 108–130

Nott, H. and Combs, G.F. (1967). *Feedstuffs*, **39**(41), 21–22

Roan, S. (1991). *Bio-economic Models for the Simulation of the Production and Management of the Growing Pigs and Sows.* PhD. Thesis, University of Edinburgh

Yutste, P., Longstaff, M.A., McNab, J.M. and McCorquodale C. (1991). *Animal Feed Science and Technology*, **35**, 289–300.

Zaalberg, J. (1989). *Journal of the Association of Official Analytical Chemists*, **72**, 34–37

IV

Feed Compounding

9

THE IMPACT OF CAP REFORM AND GATT ON THE UK ANIMAL FEED COMPOUNDING INDUSTRY

R.W. DEAN

Dean Agricultural Associates, 16, St Georges Lodge,
London N10 3TE

Introduction: the general background

While negotiations on both the Uruguay round of the General Agreement on Tariffs and Trade (GATT) and the future of the CAP are still in the balance, it is difficult to predict the long term effects on British agriculture; its total sales, its structure or the volume of economic resources - land, labour and capital - it will continue to command. It is commensurably more difficult to predict the effects on the industries supplying farmers with inputs. This business, currently worth over £7,000 million a year in the UK, is dominated by the supply of animal feed.

Table 9.1 shows that, in 1990, British farmers spent over £3,000 million on purchasing animal feed, not including feedingstuffs grown and used on their own farms. Two-thirds of these purchases by volume and over 70 per cent by value consisted of compound animal feeds. Some indicators of the changing structure of the compound feed industry in the UK are to be found in Table 9.2.

Changes both in progress and in prospect in agriculture will have a direct and significant effect on the agricultural supply trade, including the compound feed industry. A decade ago, the trade's watchword might have been "high input, high output".

The decision taken by the post-war Labour government, that the national interest required a strong farm sector, resulted in steps to continue and to enhance the development of UK agriculture initiated during the war. This represented a major policy shift. Since the abolition of the corn laws in 1869, UK agricultural policy, if it could be described as such, might best be described as "benign neglect". Indeed, one President of the Board of Agriculture, the precursor to MAFF, proclaimed that his own job was "to preside over the demise of British agriculture and to give it decent burial".

The economic philosophy which engendered this neglect of UK agriculture was a product of the high summer of British imperialism in the late nineteenth century.

Britain's future was to be the workshop of the world. This required cheap labour. To ensure cheap labour required cheap food. Britain's own farmers could not supply the burgeoning population of the British Isles at prices which would allow wages to remain low and competitive with the looming economic power of Germany and in particular, the United States.

Table 9.1. VOLUME AND COSTS OF ANIMAL FEEDS IN THE U.K. IN 1990

Type of feed	*Volume*(m tonnes)
Compound feeds	
Cattle	3.9
Calves	0.3
Pigs	2.3
Poultry	3.8
Others	0.8
Total Compound Feeds	11.1
Other High Energy Feeds	4.6
Low Energy Feeds	0.7
Total Feed Purchased	16.4
Compound Share by Volume	67.7%
	Value (£m)
Total Feed Purchased	3,017
Compounds	2,116
Straights	722
Others	179
Compound share by Value	70.1%

From MAFF (1991).

If farmers could not deliver the goods, their colleagues in the dominions and the other "new" countries could. With their enormous areas of cultivable land and low production costs, plus the development of the refrigerated vessel able to carry perishable foods halfway around the world, Britain could be supplied from overseas at a fraction of the cost of home-produced supplies. The only agricultural commodities which British farmers could produced competitively were those whose highly perishable or fragile nature, such as milk and eggs, meant that they could not be easily sourced from overseas.

Table 9.2. KEY FACTS ABOUT THE U.K. FEED COMPOUNDING INDUSTRY

Indicator	*1979*	*1983*	*1989*
Output ('000t)	11,600	12,234	10,619
Output (£m)	1,771	2,132	1,974
Number of Companies	407	383	327
Number of Mills	509	478	364
Employment ('000)	20,200	17,200	12,000
Direct Manufacturing Costs (£m)	1,496	1,762	1,598
Overheads (£m)	66	120	90
Wages and Salaries (£m)	96	130	143
Operating profit (£m)	129	105	141
Operating Margin (%)	7.2	4.9	7.1

From CSO (1991) and Dean, R.W. (unpublished data)

The origins of agricultural protectionism

Long-term UK policy has contrasted starkly with that of most European countries. In France and the emerging Germany, the rôle of agriculture was seen, not only to supply food but also to secure social and military objectives. A long, apparently never- ending series of land wars emphasized the ability to source most basic foods from national territory. The countryside also provided a ready source of manpower. The experience of blockade, as practised by the British Navy, did little to dissuade continental politicians from this view.

There was thus no question of allowing agriculture to decline in the face of North American or Australasian competition. Domestic agriculture was protected by import duties and other tariffs and internal farm prices heavily supported. The agriculture sector thus remained much larger during the first half of the present century than would have been the case had it been exposed to the sort of competition experienced by UK farmers.

In Europe, a higher proportion of available land and labour has been and continues to be utilized by agriculture. National policy has deliberately sought to create conditions allowing the farm sector to secure a higher proportion of national economic resources through taxation, monetary measures or by forcing consumers to pay higher prices than those prevailing on world markets.

An appreciation of this is amongst the keys to this chapter. Measures which reduce or remove the protective elements of the CAP, whether in the form of subsidies or in administered prices for farm products, affect the size of the agricultural sector and, in particular, its buying power. The latter includes the £7 billion plus spent by UK farmers each year with the supply trade, including compound feed manufacturers.

The CAP versus GATT

In theory, the CAP almost completely insulates European farming from world markets. Farmers are induced by an artificially maintained price level to produce basic and not-so-basic foodstuffs to the point of self-sufficiency. These prices are extracted from the consumer in a number of ways.

Most important is Community preference. In practice, this means that Community farmers get first opportunity to supply domestic consumers. Variable levies on imported alternatives, set at a level designed to bring their prices up to the level thought desirable for EC producers (the threshold price) ensure that, in practice, most imported agricultural products are un-competitive in Community markets.

If domestic EC prices fell below a designated level such as intervention, produce is withdrawn from the market and stored at Community expense until conditions improve. Examples of this include cereals, beef, butter and skim milk powder. Private storage is available for other commodities. (The difference between the threshold price and the intervention price may be seen as the cost of Community preference). This system engendered considerable opposition in world trading circles because of the way that it excluded American and other exporters of agricultural produce from EC markets.

During the 1962 Kennedy GATT round, it became clear that, if concessions were not made to the US and others, they might seek to drag agriculture into the negotiations. To ensure that the less desirable aspects of the CAP were accepted, if not embraced, by Europe's trading partners, the tariffs on a number of products imported into the EC from the US and certain other countries were reduced to zero and bound as such under GATT. These included certain oil-seeds and their residues together with a number of food manufacturing by-products used as ingredients by the animal feed industry.

The nub of the current dispute

The present tension between GATT and the EC arises because of one important fact. Agricultural production in the Community has been largely supply-led. Prices have been managed at levels calculated to secure agricultural self-sufficiency. The fact that surplus production has emerged is partly due to the open- ended nature of many of the CAP's support régimes. But the CAP has also ignored the importance of technical progress. In this respect, the supply trade has played a significant part.

Functioning in a highly competitive market, the trade has had every motivation to improve the quality and performance of the products it sells to the farmer. In a market where every litre of milk produced was either consumed, stored in intervention as butter or skim, or exported onto world markets at EC tax-payers' expense, the winners in the compound feed manufacturing industry were those able to deliver dairy feeds promising higher milk production most efficiently. Where surplus grain could be stored or exported at tax-payers' expense, there was every incentive to fertilizer and agrochemical manufacturers to develop products promising the producer higher yields per hectare. The phrase "high input, high output" began to look suspiciously like "infinite input, infinite output".

The emergence of large volumes of subsidized EC farm products onto world markets has increased US, Australasian and other agricultural producers' appetite for a fight. Export restitutions - spelt "subsidies" in Washington DC and elsewhere - are at the root of the dispute between GATT and the EC. The US administration and the Cairns group maintain that the sole reason for the EC's emergence as a major exporter of farm products is the subsidies available to exporters. The EC has no natural competitive advantage in products such as grain, meat or dairy products; rather the reverse. Thus, markets now regularly supplied by the EC have been won unfairly, contrary to the liberal, free and fair-trading philosophy enshrined in GATT.

This makes nonsense of the claim, much touted in Brussels, that CAP reform is a purely internal matter and has nothing to do with GATT. The current dispute would not have arisen, at least to the same Armageddon-invoking extent, had not the EC emerged as such a force in world agricultural markets to the discomfort of its competitors. In that it has done so, it is the CAP, in encouraging an artificial allocation of economic resource to EC agriculture, which distorts the pattern of world food production.

The opposing positions

Up to the end of November 1991, the US position had not essentially changed. This was that, over a period of ten years, three-quarters of all farm subsidies and 90 per cent of all export restitutions should be abolished. The Cairns Group, consisting of Australasia, Canada and a number of South American countries has not, in essential matters, deviated from that of the US.

The EC position, established during a marathon negotiating session in October/November 1990, was that 30 per cent of its farm subsidies should disappear over a five year period. This was attacked as provocatively disingenuous by the EC's critics who argued that half the reduction was already provided for under other EC arrangements. Despite appearing to promise a 30 per cent cut in subsidies over five years, all the Community was actually offering the GATT round was a 30 per cent cut over ten years between 1986-96 - a very minor reduction.

Despite protestations to the contrary, it has become clear that the MacSharry proposals, notably those related to cereals, are a key element in the overall package which the EC will finally lay on the table as part of its contribution to the GATT negotiations. As far as they are known at present, the measures most likely to affect the compound feed industry are as follows:

1. For cereals, a sharp cut in EC support prices with the target price being set at ECU 100/tonne, and the threshold and intervention prices ten per cent higher and ten per cent lower respectively. This would give an intervention price equivalent to about £63 a tonne.

2. Abolition of the Maximum Guaranteed Quantity and the basic co-responsibility levy for cereals.

3. Grain farmers would be compensated for loss of income but on a modulated basis. This would mean that compensation would be paid for income losses subject to their agreement to set aside designated areas of land. It would also be paid at a rate diminishing in proportion to the size of the holding.

4. A similar arrangement to 3 above is proposed for oilseed and protein crops.

5. A further cut in milk quotas is proposed which the Community expects to amount to a net reduction of 3 per cent. Small producers with less than 200,000 kg output will be exempt. Their share of the overall cut will be assumed by larger producers.

6. Small to medium sized producers whose quotas were reduced would be compensated by a headage payment for the first fifteen cows with the proviso that stocking rates do not exceed more than one livestock unit per forage hectare.

7. The basic milk co-responsibility levy will be abolished.

8. Intervention prices for butter and skim will be reduced, resulting in a fall in the Intervention Milk Price Equivalent.

9. Beef intervention prices will fall by 15 per cent and restricted to low quality material. Compensation will be paid on a restricted number of animals and on the basis of one livestock unit per forage hectare.

10. For sheep, an additional premium will be paid to reduce stocking rates, limited as to the number of ewes on which it is payable. Maximum stocking densities would also be prescribed for different areas.

What is far from clear is whether the US and its allies will see these as real cuts in the level of EC farm subsidies, likely to result in significant reductions in the volume of surplus European farm produce being dumped on world markets.

Implications for compounders

Compounders sit, in a sense, at the centre of a web being woven around the CAP and GATT. They are major purchasers of, amongst other materials, grains, oilseeds and proteins. They also sell compound animal feed to producers of, not only livestock products such as milk, beef and sheepmeat which will be directly affected by elements of Mr MacSharry's proposals, but also to pig, poultry and egg producers whose costs of production are heavily influenced by the CAP.

Table 9.3 shows the livestock sector's gross revenues and their spending on compound feeds during the 1980's. Spending on compounds accounts for a significant proportion of producers' gross revenues in the pig and poultry sector. The proportion is less in the cattle sector and was lower in 1990 than in 1980 as a result of the introduction of milk quotas. This reflects the increasing part played in total cattle feed by producers' own resources such as silage and other forage crops. Higher compound usage in the sheep sector reflects the long expansion of the industry since Peter Walker brought back the sheepmeat régime in 1980.

The term "Compounds Ratio" can be interpreted in a number ofways. It may be viewed as a straight forward measure of margin over concentrates. It is also an indicator of the proportion of feed supplied by the compound feed industry compared to other sources whether home-grown, home-mixed or supplied as part of an integrated livestock production process. It also measures the extent to which compound costs pre-empt other direct and indirect manufacturing costs - including profit - and that an undue rise in the ratio may presage action taken to reduce it either in the short or longer term.

Table 9.3. LIVESTOCK PRODUCERS GROSS REVENUE AND SPEND ON COMPOUND FEEDS IN 1980 AND 1990

	1980	*1990*
	(£m)	
Cattle and calves	1,544	1,946
Milk	1,960	2,804
Cattle Sector Total	3,504	4,750
Cattle Compound Spend	649	707
Cattle Compound Ratio (%)	*18.5*	*14.9*
Pigs	780	1,033
Pig Compound Spend	323	468
Pigs Compound Ratio (%)	*41.4*	*45.3*
Poultry	508	894
Eggs	489	544
Poultry Sector Total	997	1,438
Poultry Compound Ratio (%)	*49.8*	*55.8*
Sheep and Lambs	448	981
Wool	36	50
Sheep Sector Total	484	1.031
Sheep Compound Spend	16	81
Sheep Compound Ratio (%)	*3.3*	*7.9*

From CSO (1991) and MAFF (1991)

At this stage, a number of questions will start to suggest themselves to compounders:

1. What will be the effect of GATT/CAP reform on the proportion of cereals, proteins and other raw materials in their ingredient mix?

2. At what price can that mix be profitably delivered to UK Livestock Farmers?

3. How much feed will livestock farmers be prepared to purchase from UK compounders, given that the prices they receive for their livestock and livestock products may be affected?

4. What, if any, will be the reactions of consumers?

Compound feed supply - The raw material matrix

The importance of raw material costs in compound feed manufacture is axiomatic. Available data show that, in the ten years 1979-89, compounders' primary production costs as a percentage of total sales value averaged 82.6 per cent (range 81.0 - 84.5%; s.d 0.90). Primary production costs are defined as the cost of raw materials, fuels and power, packaging and small tools and materials. Occasional data suggest that over 90 per cent of primary production costs consist of raw materials (The Department of Trade and Industry, 1984).

Table 9.4. RAW MATERIAL INCLUSION IN EEC COMPOUND FEEDS

Ingredient	*1975*	*1988*	*1975*	*1988*
	m tonnes		% total	
Cereals	25.7	30.6	44.2	30.7
Tapioca	2.3	6.3	4.0	6.3
Industrial By-products	10.4	16.5	17.9	16.5
Oils and Fats	0.6	1.6	1.0	1.6
Oil-cake and Meals	12.5	25.8	21.5	25.9
Animal Meals	1.7	2.5	2.9	2.5
Milk Products	1.2	1.1	2.1	1.1
Dried forage	1.0	2.6	1.7	2.6
Minerals, Additives	0.8	2.0	1.4	2.0
All Others	1.9	10.7	3.3	10.7
Total	58.1	99.7	100	100

from EC Feedingstufs Committee (personal communication)

The proposal for a major cut in cereal support prices will highlight the long-standing controversy between EC farmers and compounders over the volume of cereals used in compound animal feed manufacture. Table 9.4 shows use of major raw material groups in 1975 and 1988. If the 1975 cereals inclusion rate is applied to the larger 1988 tonnage, the "loss" to cereals between the two years is equivalent to over 13 million tonnes. Part of this volume was replaced with oil-cake and meals as well as tapioca. But there was also a large increase in the proportion and variety of other materials used.

The debate over substitutes

The debate between compounders and the cereals lobby, with some elements of DGVI behaving in a less than disinterested manner, has raged around compounders' access to these materials, many of which are processing by-products originating in the food industry. The cereals lobby contends that the latter have no production costs. They can thus be priced in a way designed to force cereals out of formulations. This contention has led directly to demands that the EC should seek to unbind the zero tariffs on products like maize gluten, citrus and sugar beet pulp and other food industry derivatives.

Compounders regard this argument as fallacious on three grounds. Firstly, alternatives to cereals give compounders additional flexibility in formulating feeds. With cereal prices set by the annual farm price determination in Brussels, cereal replacers constitute the only deterrent to grain prices being set solely with reference to the perceived income needs of cereal producers.

Secondly, access to cereal replacers provides a balance to the disparate support régimes of the CAP. EC grain prices are influenced by the existence of intervention, export restitutions and the whole panoply of a "heavy" CAP régime. In contrast, cereal consumers such as pigs and poultry have "light" régimes. Their prices reflect the interaction of supply and demand.

Compounders' free access to cereal replacers goes some way to correcting this imbalance. One way of demonstrating this fact is that gross margins for cattle and, in particular, dairy feeds are higher than for those of pig and poultry feeds. Gross margins defined as delivered cost of feed less raw material cost. One recent publication suggests that while cattle feed production may yield gross margins of about £40 a tonne, pig feed margins are £8 lower and poultry feed margins may be as low as £24 on average (Dean, 1991).

A third weakness in the cereal lobby's claim against compounders' free access to cereal replacers is the assumption, implicit in their argument, that all raw materials are perfectly substituted. This contention is untenable on technical grounds. The high fibre content of many cereal replacers makes them more suitable for ruminant rations than for monogastrics. The validity of this statement is demonstrated by the much lower cereal inclusion rates in dairy feeds compared to those for pigs and poultry.

None of these arguments are invalidated by reduction of the cereals intervention price to £63 a tonne, compared to the present buying-in price for feed grains of £117. But before compounders contemplate grain prices 40 per cent below those currently prevalent, two other aspects need to be explored.

Prospects for compounders' use of cereals

The proposed CAP reforms seek to make cereal growers more responsive to the market and less dependent on the CAP's various support mechanisms. This, in part, is the objective of reducing the intervention price for grain from its present level to one of around £63 a tonne. A distinction must be made, however, between the intended function of intervention as a floor price and what it has become in recent years - a significant market in its own right.

It is likely that a market price for grain will emerge which lies between £63 and the market price in 1991. Grain production is likely to fall, but not to such an extent that EC supply and demand will balance, particularly when quality aspects are taken into account. On the other hand, most farmers will receive compensation for lost income via compulsory set-aside. The market price thus seems likely to reflect average direct production costs. In the UK, recent data suggest these are about £85 a tonne.

From compounders' viewpoint, will the prospect of, say, £90 a tonne delivered to the mill, mean a rush to grain? Two factors will influence this decision. The first is the technical preference. Grain has traditionally been regarded as the raw material of preference for technical reasons. It must seriously be questioned whether this is still so. Recently constructed or refurbished mills have been designed to handle a wide range of raw materials. Undoubtedly, some older mills could reduce production costs by using more cereals in preference to other materials but, in general, this is unlikely to be the case.

The second factor dictating raw material use is, as always, price. Where alternatives to cereals are concerned, this question comes down to the price of industrial by-products. Their prices may fall to match the lower price of cereals. However, the supply of by-products is not infinite. Feed ingredient demand is rising world-wide, especially in the emerging countries of the Pacific Rim where land and population pressures restrict the potential for additional feed grain production. The price of cereal replacers may thus start to reflect a genuine market rather than simply a price required for market clearance and determined largely by grain prices.

What about other raw materials such as oilseeds? The Commission thinks that the current relationship of cereal prices to those of oilseeds is too close and tends to favour inclusion of the latter at the expense of the former, perhaps to the extent of four million tonnes a year across the Community.

Cereal and protein prices have tended to move closer in recent years. But the Commission's proposition, that there is a "proper" relationship between grain and protein prices does not bear close scrutiny. Oilseeds are not pure protein and will enter formulations as an energy source if their price falls sufficiently relative to cereals. Furthermore, recent changes in market requirements for meat mean that higher protein levels are required to support lean meat deposition and lower energy levels to defend against fat. The concept of a "proper" price for proteins is one that compounders should not hurry to accept.

Overall, it seems reasonable to expect a 15 per cent cut in the average cost of raw materials, as a result of falls in the price of grain and other materials. In 1990, it is estimated that UK compounders purchased 11.3 million tonnes of raw materials valued at £1,400 million or an average of £124 a tonne. On this basis, compound feed manufacturers might expect raw material costs of £1,200 million at 1990 prices for raw materials post-GATT and post-CAP reform, a saving of £200 million.

Will compounders use more grain in rations which depend significantly on cereal replacers such as dairy and other cattle diets? Could we see, for example, a resurgence of beef production systems based on barley as in the early 1970's?

The industry's understanding of how to utilize the high fibre levels present in some of the cereal replacers has increased greatly in recent years. The choice between cereals and cereal replacers will partly reflect the premium charged for the former. If this rises to the point where it cancels out the technical disadvantages inherent in cereal replacers, then compounders will switch to the latter. But improvements in compounders' ability to utilize cereal replacers in preference to cereals means that the size of the premium required to induce this switch has been decreasing. It is thus unlikely that there will be a major switch back into cereals by manufacturers of ruminant feeds.

The effect on consumers - primary demand

The second part of the equation relates to the demand for compound feed. Compounders must examine two aspects with regard to the demand for their products. The first concerns demand for compound feed which is related to the structure of the livestock industries they supply - the so-called primary demand for compound feed.

Here, the social and environmental aims of Mr MacSharry's proposals are central to the problem faced by compounders. The proposals are aimed not only at cutting the cost of the CAP in financial and political terms but also at arresting - and reversing - intensification of European agriculture.

Environmental considerations are at the root of this part of CAP reform. European farms are seen as becoming progressively larger, making more intensive use of brought-in inputs with object of increasing output. As part of the process of reducing costly surpluses these structural factors must, therefore, be penalized. The EC "farm" must become smaller, use inputs less intensively and revert to being a part of the rural landscape rather than dominating it.

This strikes at the root of the type of farming, including livestock farming, supported by the agricultural supply industry. It explains in part why Mr MacSharry's plans have been greeted with such dismay by the feed industry as well as other parts of the supply trade.

One of the most objectionable aspects of the MacSharry reform plans is the suggestion that the proposed net 3 per cent cut in the volume of milk quotas should exempt those producers whose output is less than 200,000 kg. In the UK, this equates to a herd of about 40 cows. A correspondingly larger cut in quotas would affect those producers with over 200,000 kg production each year.

The first 40 cows would also be the limit for payment of the dairy cow premium, payable provided that stocking rates are under 2 livestock units/hectare (1.4 in the less favoured areas). This will hit the UK, where average herd size is more than twice the proposed cut-off level, relatively hard. The proposals, which are also designed to encourage a more extensive pattern of milk production would strike disproportionately at compound feed manufacturers. Similar considerations apply to beef and sheep production.

The effect on consumers - secondary demand

The second, and perhaps less intimidating question to be addressed by compounders is the effect of CAP reform and GATT on consumer prices for livestock products and the volume of such products consumed. This is referred to secondary demand for compound feed.

Much of the debate over how the EC supports its farmers has focused on how high levels of managed farm prices influence consumer prices for these products. The corollary is that, if ex-farm prices were lower, then so might be prices in the shops, thus raising consumption.

There are two possible routes towards lower prices for livestock products. The first is direct, in that the ex-farm price is supported at first-hand by the CAP. Such products include milk and dairy products, and beef. Alternatively, the route may be indirect, as in the case of pigmeat, poultrymeat and eggs. Here, ex-farm price is largely driven by production costs, especially feed. For example, up to 65 per cent of direct production costs for pork can be attributed to feed costs. Similarly, of a total of 115.5 pence for producing a saleable broiler at the end of 1990, 80.6 pence (70 per cent) was accounted for by feed.

For compounders as well as livestock producers, one of the most disturbing aspects of the past two decades has been the relatively static pattern of consumption of some livestock products and the decline in the consumption of others. Table 9.5 shows household consumption of selected livestock products in Great Britain in 1969 and 1989, the latest year for which data are publicly available. The data exclude consumption outside the home and for manufacturing, and therefore are an underestimate of total supplies of livestock products moving into consumption. (No up-to-date data for total consumption are available for 1988 and 1989 owing to a dispute between MAFF and other government departments on the presentation of data for total consumption of foodstuffs either at home or elsewhere).

There is evidence which suggests that the process of bringing UK food prices into line with those in the original six members of the EEC did cause some decline in the consumption of products most affected. This was because prices rose quicker than the level of prices overall - in other words in real terms. However, there is further evidence which suggests that other factors have an equal if not greater effect on the consumption of livestock products. These may be consolidated under four headings; dietary considerations, competition, social change and "welfarism".

Consumers' perception of the effect of diet on human health has increased markedly over the past twenty years. Eggs were an early victim of this phenomenon. Milk and dairy products also suffered from the association of unsaturated fatty acids with cardiovascular disease. The high fat content attributed to red meats has put them at a disadvantage compared to poultry (see Table 9.4). Successive health "scares", whether occasioned by salmonella or BSE, now have a greater effect on consumption and for a longer period than would have once been the case.

The range of products competing for a share of the consumer's purse has expanded rapidly over the past two decades. Where milk and butter would once have formed the basis of a snack, a wide range of packaged products has largely displaced the glass of milk and the slice of bread and butter as a quick "filler" in between meals.

Social change, notably the increased number of women going out to work, has diminished the home as a place of casual eating in which basic foods such as milk, dairy products and eggs figured significantly. Consumption of eggs, bacon and sausage has fallen as working women increasingly decline to prepare the Great British Breakfast.

Eating "correctly" from an ideological point-of-view has turned many consumers off meat and eggs. Such factors include the anti- factory farming lobby or straight vegetarianism, concern about how dairy cows are fed (the meat and bonemeal/BSE effect) or even the consumption of meat from animals fed on grain which could have been used in

Table 9.5. U.K. HOUSEHOLD CONSUMPTION OF LIVESTOCK PRODUCTS

Product	*1969*	*1989*	*% change*
	kg/head/week		
Total Meat	25.8	22.9	-11.2
Beef and Veal	7.7	6.0	-22.0
Mutton and Lamb	5.3	3.0	-43.4
Pork	2.8	3.2	+14.3
Bacon and Ham	5.1	3.4	-33.3
Poultry	4.9	7.3	+49.0
Liquid milk and Cream	4.9	3.9	-20.4
Butter	6.1	1.8	-70.5
Cheese	3.5	4.1	+17.5
Eggs (number)	4.6	2.3	-50.0

From MAFF (1970, 1990)

the third world.

Every commentator will have a view on which of these influences is most important. But, taken together, non-price factors are likely to assume increasing importance as determinants of the consumption of livestock products.

It is interesting to note that, during the debate over GATT and CAP reform, virtually no attention has been paid to the possibility that lower food prices might result in higher consumption and thus that lost revenue through lower prices might be made up by higher volumes.

Empirical evidence suggests that the income elasticity of demand for food in general is low, although greater for some individual high value products. Similarly, there is little to suggest that livestock products enjoy a significant price elasticity. It is, in the author's view, extremely unlikely that the prospect of lower prices leading to increased consumption will be realized (The income elasticity of most food products appears to be <1. The overall price elasticity of foods varies with the type of food under discussion but would also appear in most cases to be <1).

This view is reinforced by the increasingly concentrated nature of the UK food retailing sector. It is questionable whether any reductions in producers' first-hand prices would be passed on to consumers in any significant degree, even though the high street chains will be quick to demand lower supply prices for products where production costs are known to have fallen as a result of changes in the CAP or GATT.

The conclusion may be drawn that livestock producers' gross incomes in terms of receipts from the market place are likely to fall. This takes no account of additional income which may be received as a result of measures to compensate producers under CAP reform measures now under discussion.

An overall assessment

The highly interactive nature of the factors influencing the future shape of the compound feed industry make it very difficult to predict what the industry will look like in ten years. Certainly, one is chary of making the sort of prediction essayed by the head of one major player in the industry (M. Halluitte, President of Rhône Poulenc Animal Nutrition, addressing the FEFAC Congress in Paris in May, 1991) when he said that, by the end of the century, half the existing feed mills in Western Europe would have disappeared .

The size of the animal feed industry at any time will be determined by the profits stream stemming from its activities. These must be large enough to constitute an adequate return on funds employed in the industry.

The combined effects of CAP reform and GATT seem certain to induce a significant downwards shift in the cost of the compound feed industry's raw materials, its major cost component. This will reflect falls in the support price of grain which are likely to result in market prices of about 20 per cent lower than currently. The price of the so-called cereal replacers is likely to fall in sympathy, while protein prices will also decline although not by so much as in the case of grain. This assumes that the EC will fail in its attempt to win the right of "rebalancing" and that oilseeds and oilmeal will continue to enter the Community duty and levy-free.

What is far less certain is the overall effect upon the industry's gross revenues derived from the volume of compounds sold, and the price charged for them.

Gross Margin is an important economic factor in the compound feed industry. It is defined as the price received from the sales of compound feed less the cost of raw materials. For practical purposes, the cost of energy, packaging, small tools and materials are usually deducted at this stage, but ingredients still account for over 90 per cent of direct manufacturing costs.

The gross margin is available to pay labour costs and overheads, and to meet financial obligations such as depreciation and interest charges. What is left is pre-tax profit. With the exception of the latter, all these items may be regarded as fixed costs.

If we assume that (i) because of lower support prices and the buyers' perception of lower livestock production cost, the first- hand price of livestock products declined, overall, by fifteen percent, and (ii) that there is no significant increase in consumption in livestock products, then there will be a consequent fall in livestock producers' gross revenues of the same proportion.

From Table 9.3 it can be seen that the total gross revenue of livestock producers for 1990 amounted to £8.25 billion. From this, they spent £2.06 billion on compound feed, a compound ratio of about 25 per cent. A fall in livestock producers' gross revenue to the extent discussed would, assuming that the compounds ratio remains the same, result in a fall in compounders' revenue of over £300 million. The assumption that the compounds ratio remains unchanged at a constant 25 per cent takes no account of the possibility that CAP reform measures in particular, aimed at a less intensive production, might reduce the compounds ratio to a lower level. Each percentage point reduction in the compounds ratio would result in a further loss of revenue, based on 1990 data, of £82.5 million.

The key to the effect on compounders' gross margin will lie in the behaviour of raw material costs. These will undoubtedly fall as a result of the proposed CAP and GATT reforms. In order to continue to finance the industry's present level of fixed costs, they would have to fall by a greater percentage, as shown in Table 9.6.

Table 9.6. EFFECTS OF CHANGING GROSS MARGINS ON CASH FLOW OF THE U.K. COMPOUND FEED INDUSTRY

	Current (1990)	*Post-CAP/GATT*	*% Change*
	£billion		
Gross Revenue	2.060	1.750	-15.0
Direct Costs	1.670	1.419	-10.0
Gross Margin	0.390	0.331	-15.0
Staff Costs	0.142	0.142	n.c.
Overheads	0.100	0.100	n.c.
Operating Profit	0.148	0.089	-39.8

From Dean (1992)

The near-40 per cent contraction in operating profit, after paying fixed costs, highlights the importance of the gross margin to the compound feed industry as a source of funding other fixed and financial charges. From £148 million in 1990, it is predicted to fall to £89 million at 1990 prices when reductions in both gross revenue and direct manufacturing costs are taken into account. This raises the prospect that the industry will not be able to support its present level of fixed costs and financial charges in a post-CAP/GATT world.

CAP reform and GATT look set to initiate a new reduction in the industry. The numbers shown in Table 9.6 can only be indicative. The industry will have to undertake its own calculations in order to determine how livestock producers gross revenues are likely to fall and how compounders' gross revenues and costs will be affected as a result.

If gross margin declines to the extent suggested in the conservative assumptions spelled out in Table 9.6, then the industry will have to consider the extent to which it sheds fixed costs and financial charges - in other words, the extent to which it disinvests. We have no measure of the capital currently employed in the industry, nor of the depreciation or other financial charges which it incurs as a result of past capital investment.

However, in the same way as a fifteen per cent reduction in overhead and labour costs would be required to return their operating ratios to pre-CAP/GATT levels, the figures in Table 9.6 suggest that a 15 per cent reduction in manufacturing capacity might be required to bring the industry back to profitable operation. This figure will be correspondingly larger as assumptions about increasing efficiency are taken into account.

Finally, if the transition to a post-GATT/CAP reform world is to be smoothly managed, it is clear that the compound feed manufacturing industry needs to take advantage of the proposed transition periods - in order to assess the effects on the industry and to plan accordingly.

References

Central Statistical Office (1990). In *Business Monitor 1989.* HMSO, London

Central Statistical Office (1991). In *Annual Abstract of Statistics 1991.* HMSO, London

Dean, R.W. (1991). In *Eurofeed '91 - The New Order.* Lakebourne Information Services, London

Dean, R.W. (1992). In *The Feed Compounder*, **12** (2) (in press). HGM Publications, Baslow

Department of Trade and Industry (1984). *Quinquennial Purchase Survey.* HMSO, London

Ministry of Agriculture Fisheries and Food (1970). In *Household Food Consumption and Expenditure 1969.* Report of the National Food Survey Committee. HMSO, London

Ministry of Agriculture Fisheries and Food (1990). In *Household Food Consumption and Expenditure 1969.* Report of the National Food Survey Committee. HMSO, London

Ministry of Agriculture Fisheries and Food (1991). *Agriculture in the United Kingdom 1990.* HMSO, London

10

EFFECTS OF PROCESSING ON THE STABILITY OF HEAT LABILE NUTRIENTS IN ANIMAL FEEDS

J.R. PICKFORD

Tecracon Ltd., Braintree, Essex, UK

Introduction

In the manufacture of animal feeds the pressing of meals into cubes and pellets is a long established practice. It is recognised that this process causes some loss of heat labile nutrients, particularly some of the vitamins (Pickford, 1968). Adjustments are therefore made in the formulation of feeds to allow for such losses.

However, in the past three years significant changes have been made in feed mills both by modification of existing practices and the introduction of new machinery. The resulting, more severe treatment of meals through the pelleting process has raised the question of whether these generally beneficial changes in manufacturing practices could have a more destructive effect on heat sensitive feed nutrients and additives.

There are three main reasons for these changes in production processes:

1. To improve the effectiveness of the manufacturing process in producing feeds free from harmful bacteria, particularly Salmonella.

2. To improve the quality and durability of the cubes and pellets.

3. To increase the range of ingredients which can be efficiently handled and to increase the level of liquid ingredients, such as fat and molasses, which can be incorporated into the mixed meal prior to pressing.

There are various ways in which these changes in production systems can be made, but there are a number of common factors which are involved:

1. An increase in the temperature of processing.

2. An increase in the time at which the meal is held at the higher temperatures.

3. An increase in the pressure to which the meal is subjected during processing in most cases.

All of these factors, when applied in the moist conditions normally required during the pelleting process, can be considered as potentially destructive to certain nutrients and additives.

Changes in manufacturing processes

As a process modification, double pelleting has become virtually standard practice in many modern feed mills. Double pelleting is simply one press discharging pressed material into a second press directly underneath. Mixed meal is heated by steam in a standard conditioner above the first press but there is no significant increase in the conditioning time. The time increase at which the meal is held at higher temperatures between entry to the conditioner and pellet discharge from the second press is very short, 20 to 30 seconds, but there is a significant increase in the maximum temperature reached. This may be as much as 12°C higher than the temperature of around 80°C which is achieved by single pelleting.

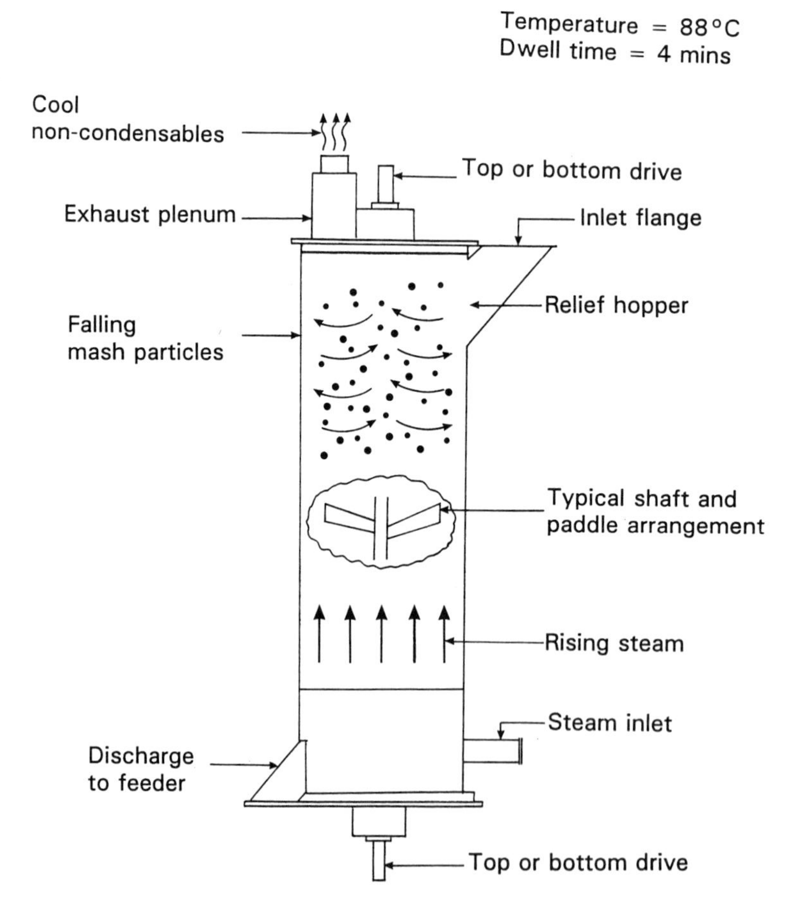

Figure 10.1. Diagrammatic representation of the structure and operation of an anaerobic pasteurising conditioning system

There are two types of new plant being used in feed mills. The first is the Anaerobic Pasteurizing Conditioning (APC) System. In this system incoming meal is treated in a vertical conditioner with steam and hot gases from a direct fired steam generating unit as shown in Figure 10.1.

The fuel used in the steam generator is natural gas, and the flue gases together with the steam pass into the conditioner. Anaerobic conditions are created in the conditioner; the dwell time of around 4 minutes and the controlled temperature rise to a maximum of 88°C are very effective in sterilising the meal. No pressure is applied during the conditioning and the strict control of temperature, together with the lower moisture increase from the use of steam and flue gases, reduce nutrient losses to a minimum.

The second type is the Expander, or to give it its full name, the Annular Gap Expander. There are a number of types of this equipment in use but the principles of construction and operation are the same. A typical Expander is shown in outline in Figure 10.2.

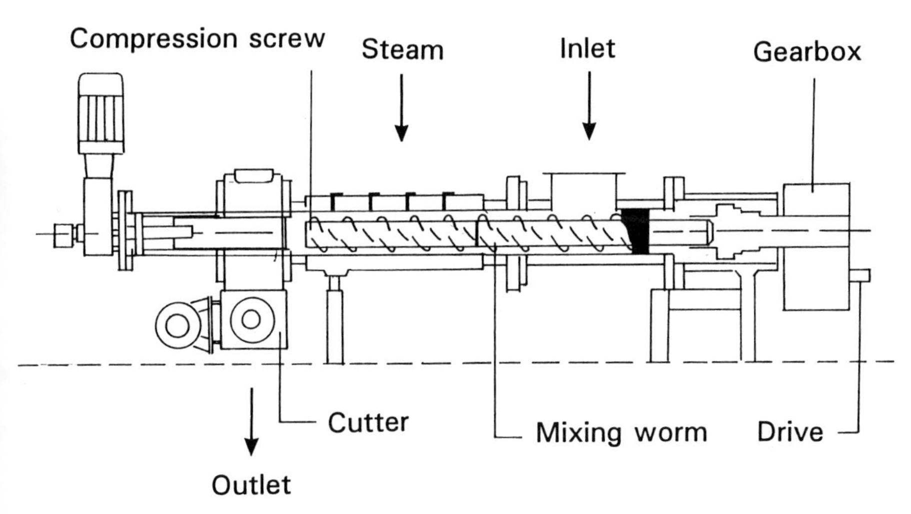

Figure 10.2. Structure of the annular gap expander

Steam conditioned meal is fed into a compression screw into which more steam is injected, and the mass is then subjected to increasing pressure and shear action and then forced through a variable exit gap. The compressed product, after reduction in particle size, is fed into a standard pelleting press as shown in the typical flow diagram of Figure 10.3.

The dwell time of feed in an Expander is from 5 to 20 seconds and the temperature reached may be as high as 145°C, but is usually in the range 130°C to 135°C. A compression pressure of the order of 35 bar is reached, but the build up of both temperature and pressure is very quick and both fall rapidly as the product is expelled through the exit gap (Figure 10.4).

An Expander is very effective in eliminating harmful bacteria, particularly Salmonella. Furthermore, it is possible to include relatively high levels of fat and other liquid ingredients prior to pelleting, and the cooking and pressure/shear action results in greater starch gelatinisation.

Expanders are single screw machines and should not be confused with the High Temperature Cooker Extruders used to produce low density pet and fish foods and in the human food industry. The processing conditions of Cooker Extruders are more severe

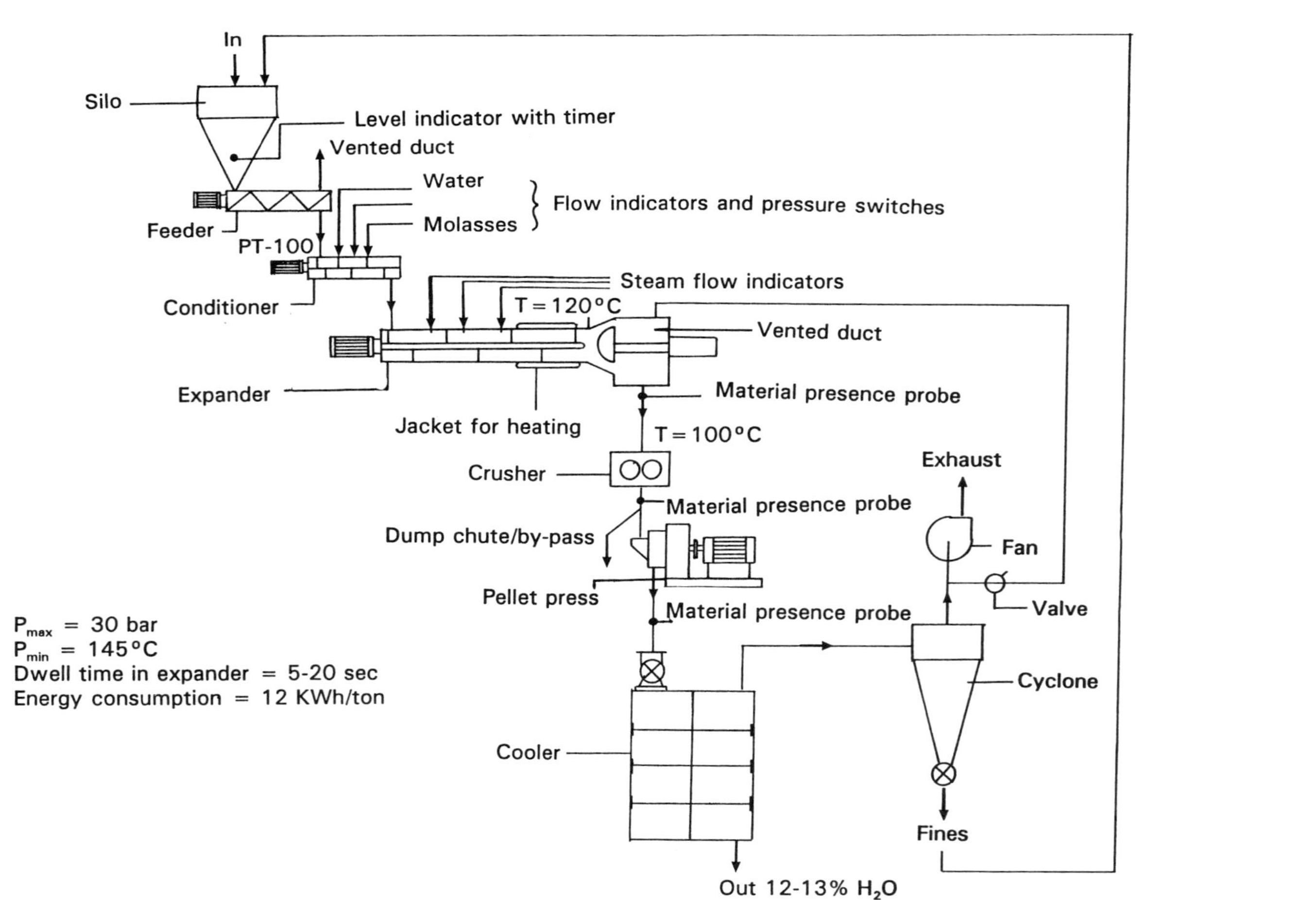

Figure 10:3: Outline of standard pelleting press system for animal feeds

than Expanders, but because they have been in use much longer there is more information available on their effects on feed nutrients. Such data may be used as a reference point in endeavouring to assess the possible effects of Expander processing.

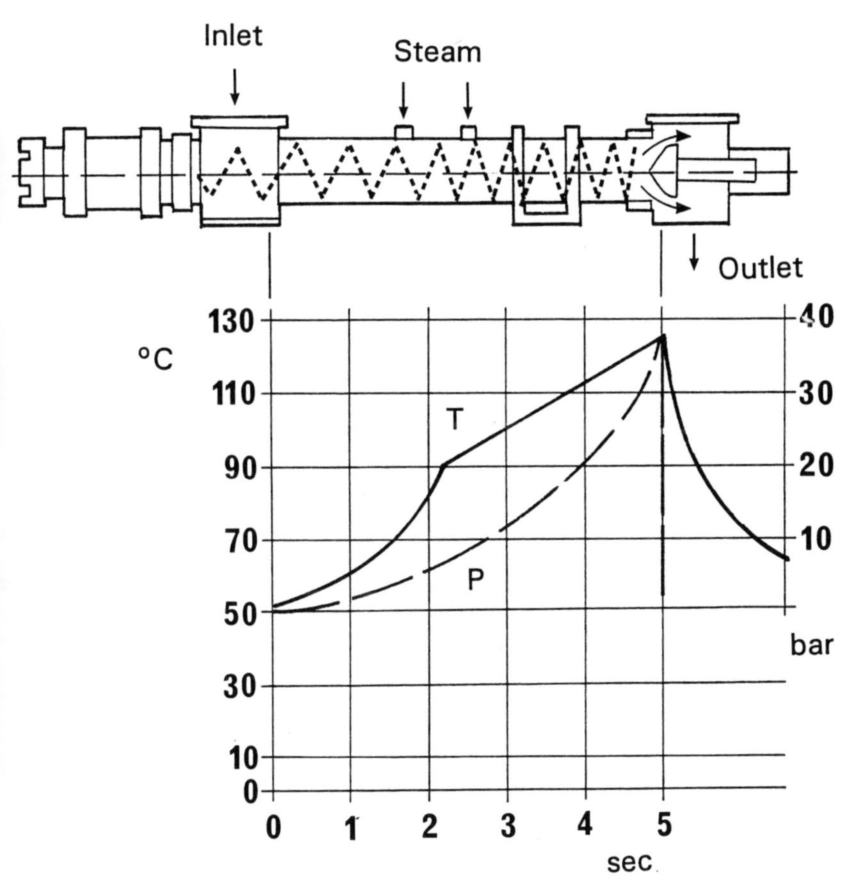

Figure 10.4. Typical build-up and fall of temperature (T) and pressure (P) through expander

The temperature/time patterns of these pre-pelleting processes are very different. Figure 10.5 shows the temperature pattern for normal short-term conditioning with single pelleting. A temperature of 80°C can be reached in the conditioner and this is increased and peaks during pelleting due to frictional heat in the dye. The total time of 15 minutes shows how long it takes for the pellets to be brought back to ambient temperature through the cooling system. The maximum temperature reached is increased by double pelleting, but the total time to return to ambient temperature will be much the same.

In order to improve liquid absorption and achieve a better elimination of bacteria, some feed mills may hold the meal at higher temperatures for several minutes as shown in Figure 10.5b. Because of the longer time at higher temperature and the high moisture levels involved, the loss of heat sensitive nutrients will be increased.

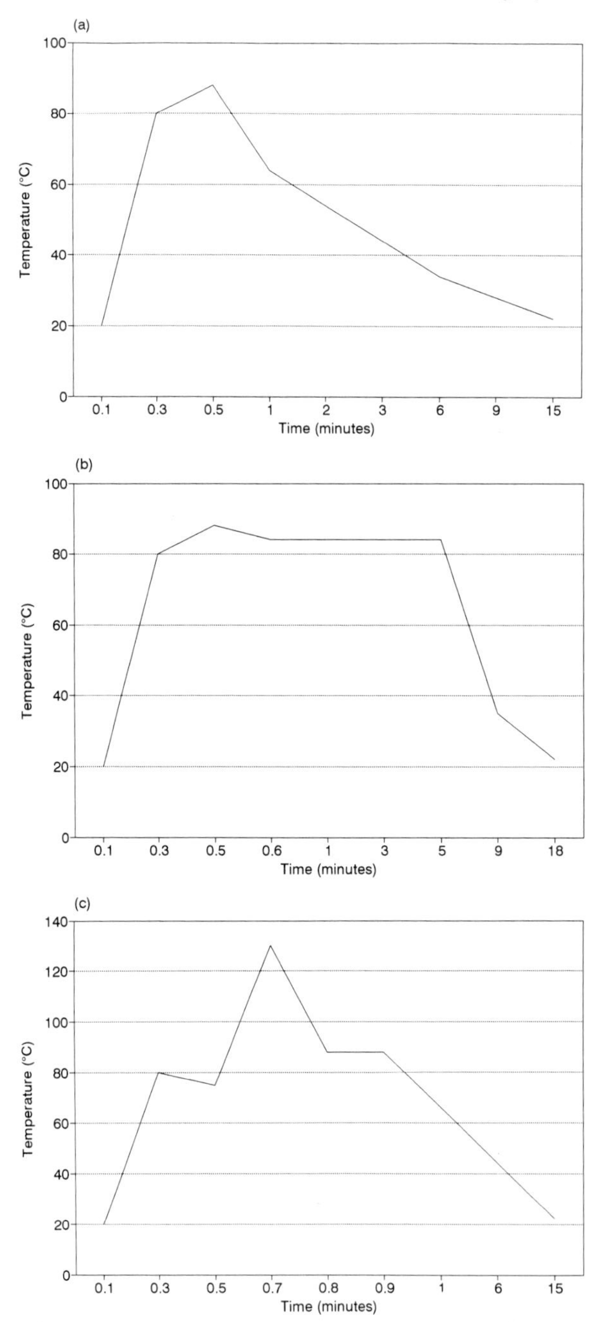

Figure 10.5. Temperature development in pre-pelleting processes. (a, Short-time conditioning without retention; b, Short-time conditioning with retention; c, Short-time conditioning with Expander)

The overall time of processing is not significantly increased when the conditioned meal goes through an Expander before pelleting, but a much higher temperature is reached, albeit for a very short time as shown in Figure 10.5c. It should be noted that the maximum temperature reached of approximately 130°C, as shown in Figure 10.5c, is often exceeded in practice due to the varying nature of feed mixes, incorrect plant operation or poor control.

Effect of processing on the stability of heat labile nutrients

VITAMINS

These developments in feed processing have increased exposure to heat together with higher moisture in some systems, both of which are factors affecting the stability of vitamins as shown in Table 10.1. It is clear that the stability of some of the important vitamins which are added to feed in synthetic form may be adversely affected by these newer processing methods.

Table 10.1. QUALITATIVE ASSESSMENT OF THE SENSITIVITY OF VITAMINS TO EXTERNAL FACTORS (KEY: -, NOT SENSITIVE TO HARDLY SENSITIVE; 0, SLIGHTLY SENSITIVE TO SENSITIVE; +, VERY SENSITIVE)

Vitamin	*Heat*	*Moisture*	*Oxygen*
A	+	0	+
D_3	0	0	+
E	-	-/0	+/0
K	0	0	0
B_1	+	0	-
B_2	-	0	-
B_6	0	0	-
B_{12}	0	0	0
Pantothenic acid	0	+	-
Niacin	-	-	-
Biotin	0	-	-
Folic acid	+	-	-
C	0	0	+

(From Schulde, 1986)

The following presents the results of vitamin assays carried out on feeds before and after processing by various systems, with all tests having been carried out under practical feed mill operating conditions (i.e. they were not controlled scientific test runs).

Tables 10.2 and 10.3 show the results of mill tests using the Anaerobic Pasteurizing Conditioning (APC) system. Although as might be expected the results are variable from one type of feed to another, it does oot appear that the APC system causes a much bigger loss of sensitive vitamins than occurs in the normal pelleting process. Although the maximum operating temperature of 88°C may be held for most of the 4 minute passage time, there is very little rise in the moisture content of the meal due to the relatively dry nature of the mixture of steam and flue gases entering the conditioner.

Table 10.2. EFFECT OF CONDITIONING AND PELLETING ON THE LEVELS OF VITAMINS IN TURKEY FEEDS. IN ALL CASES THE RESULTS ARE EXPRESSED AS A % OF THE LEVELS IN SAMPLES OF THE MEAL PRIOR TO CONDITIONING AND PELLETING

	Vitamin A	*Vitamin B_1*	*Pyridoxine*	*Folic acid*
		Average % retention		
Turkey grower	87	93	94	82
Turkey concentrate	100	108	105	106

(from Roche Products, 1991)

The inevitable assay variation for vitamins is well illustrated by the figures in Tables 10.2 and 10.3, and should be borne in mind when assessing these and other similar assay results.

Table 10.4 shows the results of two tests on feeds subjected to double pelleting. As with much of the commercial feed mill data available, basic information such as the temperature reached is not available, but it is known that the temperature rise between the first and second press is 10°C to 12°C, giving an ex-die temperature at the second press of over 90°C. Taken at face value these figures support the view that the more severe treatment from double pelleting leads to more destruction of sensitive vitamins than single pelleting.

Table 10.5 shows the effect of pelleting temperatures on Vitamin A stability. These figures are the result of a mill test on the same type of feed going through the same plant. As the process temperature rises so does the loss of Vitamin A, and it can be assumed that there would also be a higher loss of other heat sensitive vitamins with increasing temperature. The synthetic Vitamin A added to feeds is in a wax beadlet protected form, and the effectiveness of this is adversely affected by physical attrition during pelleting. Further loss during storage occurs by oxidation, and the 60-day storage tests shown in Table 10.5 indicate the accelerating rate of loss following higher pelleting temperatures.

Table 10.3. EFFECT OF APC AND NON-APC PROCESSING SYSTEMS ON LEVELS OF VITAMINS IN BROILER FEEDS. THE RESULTS ARE EXPRESSED AS A % OF THE LEVELS IN SAMPLES OF MEAL PRIOR TO CONDITIONING AND PELLETING

		Non-APC system	*APC system*
		Average % retention	
Broiler grower			
	Vitamin A	100	90
	Vitamin E	92	104
	Vitamin B_1	90	83
	Vitamin B_2	71	68
Broiler finisher			
	Vitamin A	92	84
	Vitamin E	102	92
	Vitamin B_1	83	73
	Vitamin B_2	76	68

(from Roche Products, 1991)

Table 10.4. VITAMIN LEVELS IN DAIRY FEEDS SAMPLED AFTER FIRST PRESS AND AGAIN AFTER SECOND PRESS IN A DOUBLE-PELLETING SYSTEM. ALL VALUES ARE EXPRESSED AS A % OF THE LEVELS IN THE MEAL PRIOR TO PROCESSING

	Test 1		*Test 2*	
	1st press	*2nd press*	*1st press*	*2nd press*
Vitamin A	65	71	91	65
Vitamin E	91	86	94	93
Vitamin B_1	103	61	76	64

(from Roche Products, 1991)

Table 10.5. EFFECT OF PELLETING TEMPERATURE ON THE STABILITY OF VITAMIN A. ALL VALUES ARE EXPRESSED AS A % OF THE LEVELS IN THE MEAL PRIOR TO PELLETING

	% Vitamin A recovery	
Pelleting temperature (°C)	*Ex plant*	*After 60 days*
70	96	91
80	99	89
85	90	86
90	85	75
95	88	72

Table 10.6. RANGE OF VITAMIN LOSSES DURING EXTRUSION COOKING

	% losses
Vitamin A	12 - 88
Vitamin E	7 - 86
Vitamin C	0 - 87
Vitamin B_1	6 - 62
Vitamin B_2	0 - 40
Vitamin B_6	4 - 44
Vitamin B_{12}	1 - 40
Niacin	0 - 40
Folic acid	8 - 65
Calcium pantothenate	0 - 10
Biotin	3 - 26

(from Schulde, 1986)

Although it can be seen that increasing the pelleting temperature results in the higher losses of heat labile vitamins, what is not clear is the effect of the higher pressure of double pelleting, and the much higher pressure to which the feed is subjected in expander processing. Expanders may be described as a simpler variation of and a development from the more complex Cooker Extruders such as the Wenger which have been in use in the fish and pet food industries for many years. Cooker Extruders operate at temperatures of 200°C or higher and similar exit pressures to Expanders. Schulde (1986) has listed the range of vitamin losses which can occur during extrusion cooking, and these are presented in Table 10.6. The wide range of production losses for each vitamin shows the effect of variable parameters such as raw materials and processing conditions on vitamin stability.

Feed mill tests on Expander processing show that since the Expander operates at higher temperatures the temperature of the pellets exiting the press increases and the loss of Vitamin A is greater. The figures in Table 10.7 show an increase in Vitamin A loss of 15% as the Expander temperature rises from 80°C to 120°C. Since the usual Expander exit temperature is 130°C to 135°C, the indications are that losses of approximately 25% for Vitamin A should be allowed for, and losses of other susceptible vitamins may be of a similar order. The lower Vitamin A recovery after 60 days storage follows the expected pattern of higher losses at higher processing temperatures.

Table 10.7. EFFECT OF EXPANDER TEMPERATURE ON VITAMIN A STABILITY. ALL VALUES ARE EXPRESSED AS A % OF THE LEVELS IN THE MEAL PRIOR TO PROCESSING (MOISTURE CONDITIONS IN THIS TEST WERE: PELLETS EX-PRESS 16.5%; PELLETS EX-COOLER 13.0%)

Expander exit temperature (°C)	*Pellet press exit temperature* (°C)	*% Vitamin A recovery*	
		Ex press	*After 60 days*
80	73	91	86
100	81	88	74
120	97	76	64

Roche Products (1991) have published figures for average loss for thirteen vitamins and carotenoids, all of which are important in animal feeds, following pelleting at two temperatures. These figures are the results of many vitamin assays on feeds from many different feed mills and show the serious destructive effects of higher temperature pelleting on heat sensitive vitamins. The increase in the pelleting temperature from 70°C to 90°C more than doubled the average loss for the vitamins listed (Table 10.8).

Prior to the tendency over the past three years to increase pelleting temperatures to eliminate harmful bacteria such as Salmonella, normal pelleting temperature would have been in the range 70°C to 75°C. Now the target is to increase pelleting temperature above 85°C, which is the indicated temperature for effective Salmonella kill. Indeed, 90°C is now the more general pelleting temperature achieved by double pelleting or Expander processing. That such techniques are effective in eliminating bacteria is shown by the data in Table 10.9 which are the results of a mill test for Total Viable Organisms (TVO) following pelleting at various temperatures. At each temperature a comparison was made between the counts in the meal before processing and after pelleting.

Table 10.8. AVERAGE VITAMIN LOSSES IN FEED PELLETED UNDER U.K. CONDITIONS AT TWO DIFFERENT TEMPERATURES

Vitamin	*Pelleted at 70° C*	*Pelleted at 90° C*
	Average % loss	
A	10	30
D_3	15	35
E	10	15
K	20	40
B_1	15	50
B_2	10	15
B_6	10	30
Niacin	5	10
Pantothenate	10	10
Biotin	10	35
Folic acid	20	45
B_{12}	10	25
C	40	85
Carotenoids	15	25

(from Roche Products, 1991)

Table 10.9. EFFECT OF PELLETING TEMPERATURE ON TOTAL VIABLE ORGANISM (TVO) COUNTS IN ANIMAL FEEDS

Temperature (°C)	*Meal TVO count*	*Pellet TVO count*
70	0.3×10^6	5.4×10^4
80	3.2×10^6	3.1×10^3
85	7.4×10^6	1.3×10^3
90	0.9×10^6	0.7×10^2
95	1.9×10^6	None detected

ENZYMES

The addition of enzymes such as β-gluconase and xylanase to feeds for pigs and poultry to improve the availability and digestibility of carbohydrates is becoming much more standard practice in commercial feed production, and such additions can give worthwhile returns in terms of better overall performance from feeds. It is a fact that such enzymes are very susceptible to destruction by heat, and the higher the temperature to which they are exposed during processing the greater the loss will be. Table 10.10 shows the effect of increasing pelleting temperatures on the stability of three commercial enzyme preparations. A pelleting temperature above 80°C leads to an unacceptable high activity loss in these tests on straight pelleting.

Table 10.10. EFFECT OF INCREASING PELLETING TEMPERATURE ON ENZYME STABILITY. ALL VALUES ARE EXPRESSED AS A % OF THE LEVELS IN THE MEAL PRIOR TO PELLETING

Temperature (°C)	*Product A*	*Product B*	*Product C*
		% recovery of enzyme	
65	100	99	100
70	100	97	90
75	98	85	73
80	85	55	33
85	56	37	0
90	12	2	0
95	0	0	0

The additional loss when meal is put through an Expander prior to pelleting is shown by the data in Table 10.11, the higher Expander head temperatures being reflected in higher pellet press exit temperatures. At the lowest Expander treatment of 85°C and the relatively low press temperature of 68°C little loss occurred. When the Expander head temperature was above 100°C higher losses occurred, and it should be borne in mind that Expander temperatures in practice are considerably higher and therefore more destructive than this.

In the tests presented in Tables 10.10 and 10.11 the enzyme products were in an unstabilised form. However, there are now stabilised enzymes available from commercial manufacturers and the techniques used in stabilisation can be quite effective in reducing losses, as the comparative figures for stabilised and unstabilised products in Table 10.12 show. The indications are that about 50% activity of dry stabilised enzyme preparations may be retained when processing temperatures reach 95°C, and it would be prudent to assume that this is the sort of temperature reached by present production systems whether the process involves double pelleting, Expander or the APC Conditioner.

To correct for such loss, an average doubling of the levels of the enzyme preparation added to the feed would be needed, and it may be that such a level is commercially viable. However, a safer and much more reliable method would be to add enzymes to the cooled pellets after processing. Die based enzyme slurry preparations are commercially available and provided an accurate and properly designed spraying technique is used this system is obviously more reliable.

Table 10.11. EFFECT OF TEMPERATURE ON ENZYME RECOVERY AFTER PROCESSING THROUGH AN EXPANDER AND PELLETER. ALL VALUES ARE EXPRESSED AS A % OF THE LEVELS IN THE MEAL PRIOR TO PROCESSING

Expander head	*Press*	*Product A*	*Product B*
Temperature (°C)		% recovery	
85	68	100	95
90	71	93	82
95	75	80	61
100	78	63	19
105	83	31	0
110	90	1	0
115	93	0	0

Table 10.12. EFFECT OF PELLETING TEMPERATURE ON THE STABILITY OF ENZYME PREPARATIONS. ALL VALUES ARE EXPRESSED AS A % OF THE LEVELS PRIOR TO PELLETING

		Gluconase	*Xylanase*
		% recovery of enzyme	
Stabilised preparation			
	Pelleted at 75°C	101	76
	Pelleted at 95°C	49	34
Unstabilised preparation			
	Pelleted at 75°C	44	48
	Pelleted at 95°C	12	12

PROTEINS AND AMINO ACIDS

Heat can also have a detrimental effect on proteins, the more sensitive amino acids being

Cystine, Lysine, Arginine, Threonine and Serine. Asp (1986) in considering the effect of cooker extruders on protein in feeds stated that "Heat processing is known to diminish protein nutritional value by loss of essential amino acids through the Maillard reaction or oxidation. Lysine is the most active amino acid." In a trial carried out by Asp on a Creusot Loire cooker extruder at two temperatures, losses of available lysine when compared to original uncooked material were 12% at 170°C and 44% at 210°C.

As previously indicated, the processing conditions of feed passing through an Expander followed by pelleting are not as severe as those for a cooker extruder, but the trend indication is still valid. There are few data available on the effect of Expanders on protein quality, but the results of a mill test shown in Table 10.13 illustrate that higher processing temperatures do have a detrimental, albeit small effect on amino acid digestibility. Such losses could, however, have a significant effect on the efficiency of utilisation of high density production diets.

Table 10.13. EFFECT OF TYPE AND TEMPERATURE OF PROCESSING ON AMINO ACID DIGESTIBILITY

	Amino acid digestibility (%)
Unprocessed meal	85.0
Pelleter at 70°C	85.0
Pelleter at 80°C	83.8
Expander at 70°C, Pelleter at 75°C	85.0
Expander at 95°C, Pelleter at 90°C	82.8
Expander at 110°C, Pelleter at 95°C	81.0

In spite of these potential effects of higher processing temperatures on protein and amino acid digestibility, current processing techniques do improve the quality of the feed and no adverse effects on the performance have been found. General commercial experience is that production efficiency from the feed is improved but this is seldom quantified. If there is any significant loss of protein digestibility then it is probably masked by the improvement in starch and other carbohydrate digestibility. There does, however, appear to be a need for further investigation into the effects of processing methods on amino acid availability

Summary and recommendations

Newer processing methods in animal feed production are heating feeds to higher temperatures, and in some systems are subjecting the mixed meal to higher pressures and a more severe degree of mechanical working. These methods, particularly when the moisture content of the feed during processing is increased, create conditions which are destructive to heat sensitive micro ingredients. Higher processing losses of these nutrients will then arise.

The use of enzyme preparations in feeds is becoming established and they are proving beneficial for improving the performance of pig and poultry feeds. Enzyme development will continue but it is important to recognise that enzymes are generally very susceptible to loss at higher temperatures, and as processing temperatures have risen so have the levels of enzyme loss.

Little attention seems to have been given to the possibility that higher processing temperatures in the higher moisture, higher pressure environment in which they are applied to feed may adversely affect the availability and digestibility of proteins and amino acids. It is possible that some protein damage does occur, but the effects of this on animal performance are masked by improvements in the digestibility of other nutrients.

Two practical recommendations are put forward for those concerned with animal feed production and performance:

1. In order to compensate for the destructive effects which modern higher temperature feed processing systems can have on heat sensitive vitamins and carotenoid pigments, average losses of not less than 25assumed and allowed for during manufacture of the feed.

2. Only stabilised enzyme products should be used as feed additives, and even then higher levels may be needed to maintain sufficient activity in the finished feed, particularly where high processing temperatures are used.

References

Asp, N.G. (1986). In *Extrusion Technology for the Food Industry*, pp. 16-21. Ed. C. O'Connor. Elsevier Applied Science, London

Pickford J.R. (1968). In *Proceedings of Second Nutrition Conference for Feed Manufacturers*, pp. 175-184. Eds. H. Swan and D. Lewis. J. & A. Churchill, London

Roche Products (1991). VITEC Series of Technical Leaflets Reference Numbers G3 2/1, 2/3, 2/5. F Hoffmann - La Roche Ltd, Basel

Schulde M. (1986). In *Extrusion Technology for the Food Industry*, pp. 22-34. Ed. C. O'Connor. Elsevier Applied Science, London

11

LEGISLATION AND ITS EFFECTS ON THE FEED COMPOUNDER

D.R. WILLIAMS

BOCM Silcock Ltd., Selby, North Yorkshire, YO8 7DT, UK

There have been a number of significant pieces of legislation affecting the feed compounder which have or will shortly be introduced in the UK. This chapter addresses the principal areas of the new legislation.

Medicated feeds

The introduction of the new regulations has been delayed until 1 February 1992 from the October 1991 date in the EEC Directive (90/167).

The main changes to the current regulations are in the following areas:

- *Codes of practice for manufacturers* - The changes to codes A and B are minor and include passages from the EEC additives directives concerning restrictions on combining certain types of medicinal products in the same feed. Records have to be updated daily on usages of prescription only medicinal (POM) products and their distribution in feeds.

- *POM Combination Products* - It will be an offence to incorporate a combination of POMs, except in accordance with a veterinary written direction (VWD), and then only, where in the opinion of a vet, there is no effective licensed medicinal product for the disease to be treated or for the species concerned.

  This places a responsibility on the incorporator (compounder) to ensure that the vet has justified the reason for using non-licensed combinations on the new-style VWD.

- *Veterinary Written Direction* - An amended version of the current VWD is included in the new regulations as required by the EEC Directive (90/167). A copy is appended (Appendix A). The main changes are as follows:

  (a) Valid for 31 days (current is 30 days);

  (b) Disease treated to be named;

  (c) Manufacturer has to fill in Section III detailing dates of delivery, "to be used before" date, and signature;

(d) Reasons (by the vet) for authorising incorporation by a manufacturer of an unlicensed combination of medicinal products;

(e) A faxed copy of the VWD may be sent to the incorporator/supplier in an emergency provided the original VWD is sent within 72 hours.

- *Distributors* - All distributors of PML and POM final medicated feeds who are not on A or B registers of manufacturers of medicated feed (e.g. merchants) will have to register with the Royal Pharmaceutical Society of Great Britain (RPSGB) or the Department of Agriculture for Northern Ireland (DANI) in Northern Ireland. The conditions of registration in either Category 1 or Category 2 are similar to those which apply currently to the distribution/sale or supply of medicated intermediate feeds, i.e. conformance with the relevant codes of practice, payment of fees, and subject to inspections.

- *Withdrawal periods* - The regulations introduce, for the first time, restrictions on the slaughter of animals. Where medicated feedingstuffs are administered to animals whose meat, flesh, offals or products are intended for human consumption the livestock producer/farmer shall ensure that treated animals are not slaughtered to be offered for human consumption before the end of the withdrawal period.

It is likely that RPSGB (or DANI in N. Ireland) will be the enforcing agency. The regulations are expected to come into force on 1st February 1992.

In response to a request from UKASTA for a 6 month period during which no enforcement would take place, MAFF have said that it is not possible to agree formally. However, the enforcement agency RPSGB will exercise discretion whilst compounders and distributors adapt to the new requirements.

Marketing of feeds: New UK regulations

The amending EEC Compounds Directive comes into force on 22 January 1992 and will be brought into effect in the new Feedingstuffs Regulations which had still not been published by late December 1991.

UKASTA have drafted an advisory note in consultation with MAFF which will be sent to their feed members. The following is a copy of this draft which may be amended editorially but not in substance.

NEW COMPULSORY REQUIREMENTS FOR INCLUSION ON THE STATUTORY STATEMENT:

Net Quantity

Although it has been common practice throughout the feed industry to provide, voluntarily, a declaration of weight, the Regulations now require that "the net quantity" be given. This is to be expressed in the case of solid products in units of mass, and in the case of liquid products in units of mass or volume. For example, "25 kg" or "2.001 tonnes".

It is generally understood that net quantity means the aggregate weight or volume of goods without packing. Animal feed is subject to the minimum quantity provisions of the Weights and Measures Act 1985 which means that the feed must weigh at least what is declared. Pet food, however, is subject to the average weight provisions of the Weights and Measures (Packaged Goods) Regulations 1986, and 'manufactured animal feed in biscuit or cake form prepacked in a quantity by number of 16 or less' has to be declared by number (e.g. dog biscuits).

The Local Authorities Trading Standards Association (LACOTS) have pointed out that the quantity statement must be correct. Short weight has occurred on occasions sometimes because of incorrect machinery, operator error or non-weighing of product. Any compounder in doubt should contact the local Trading Standard Department for advice.

Minimum Storage Life

This Declaration has to be expressed as follows:

(a) "Use before....." followed by the date (day, month and year) in the case of microbiologically highly perishable feedingstuffs;

(b) "Best before....." followed by the date (month and year) in the case of other feedingstuffs.

(N.B. It is unlikely that many animal feedingstuffs would be of microbiologically highly perishable nature, but pet foods with a high moisture content could fall within this category).

The Regulations state that where other provisions concerning compound feedingstuffs require an indication of minimum storage life, only the earliest date should be indicated.

It is recognised that there will be more than one shelf life for the different components of a compound feedingstuff. It is, however, considered that for the "Best before" declaration, the date used should be the same as that as the expiry date of the vitamin guarantee or the medicines guarantee whichever is the earlier.

LACOTS have pointed out that it is not an offence to sell a compound feedingstuff if the "Best before" declaration has been exceeded. The situation could be different for the "Use before" date because of the nature of the feedingstuffs concerned. Action may only be taken if, on sampling and analysis, it is found that the declaration of contents falls outside the permitted tolerances. MAFF have stated that, although selling a feed after the "Use before" date has expired is not an offence in itself, it is an offence to sell an animal feedingstuff which is unwholesome or a danger to animals and/or human health.

Batch Number or Date of Manufacture

A new requirement is that either the batch number or the date of manufacture should be given. If the former is given, this should be a unique reference number on the feed ticket or accompanying document. If the latter is given, this should be in the form of the day/month/year.

Moisture

The new Regulations require the moisture content to be declared if it exceeds:

(a) 7% in the case of milk replacer feeds and other compound feedingstuffs with a milk product content exceeding 40%;

(b) 5% in the case of mineral feedingstuffs containing no organic substances;

(c) 10% in the case of mineral feedingstuffs containing organic substances;

(d) 14% in the case of other compound feedingstuffs.

The legislation lays down the following limits of variation for moisture in compound feedingstuffs if present in excess (with the exception of dog and cat foods):

(a) 1% for declarations of 10% or more;

(b) 10% of the amount stated for declarations of 5% or more but less than 10%;

(c) 0.5% for declarations less than 5%.

If no declaration is made in the case of an "other compound feedingstuff" and on analysis the moisture content is found to be 14.1% or more, an offence will have been committed under the Regulations. On the other hand, if a declaration of 14% is made, the limits of variation will apply, i.e. no offence is committed unless on analysis a level of 15.1% or higher is found.

LACOTS have issued advice to enforcement authorities stating that, in view of the requirements on moisture content and the hygroscopic nature of feedingstuffs, sampling officers are reminded about the need to ensure that samples of feedingstuffs are placed in airtight containers to prevent moisture uptake after sampling. With particular reference to bulk feedingstuffs, it should be noted that until the legal transfer is made the feedingstuffs are the responsibility of the manufacturer.

Ash Insoluble in Hydrochloric Acid

The level of ash insoluble in hydrochloric acid shall not exceed 3.3% of the dry matter in the case of compound feedingstuffs composed mainly of rice by-products and 2.2% of the dry matter in other cases. This means that it will be an offence to sell or offer for sale any compound feedingstuff (other than that below) with more than 2.2% ash insoluble in hydrochloric acid on a dry matter basis. This is equivalent to a maximum of about 1.9% in a compound of about 12% moisture. The 2.2% level may be exceeded in the case of:

(a) Compound feedingstuffs containing authorised mineral binding agents;

(b) Mineral compound feedingstuffs;

(c) Compound feedingstuffs containing more than 50% of sugar beet chips or sugar beet pulp;

(d) Compound feedingstuffs intended for farmed fish with a fish meal content of over 15%;

provided that the level is declared as a percentage of the feedingstuff as such.

(N.B. In connection with the use of authorised mineral binding agents, it should be noted that some have to be declared i.e. bentonite and montmorillonite).

LACOTS advise that random sampling of feedingstuffs may be undertaken in order to check that this provision is being complied with. Legal advice on the definition of "composed mainly of rice by-products" is that it must contain more than 50% of the rice by-products.

Other Analytical Constituents

The major change here is the requirement for a compulsory declaration of lysine in complete pig feedingstuffs and complementary pig feedingstuffs and methionine in complete poultry feedingstuffs and complementary poultry feedingstuffs.

The declarations apply to total lysine and methionine, not "available" amino acid.

The advice to be circulated by LACOTS to all Enforcement Authorities, on the analysis for the amino acid content of feedingstuffs is to be provided as soon as it is available.

Declaration of Ingredients

There is a choice between declaring by individual ingredients or by categories.

Individual Ingredient Listing

The Regulations require that a declaration of ingredients be given from 22 January 1992. Up to the present time, a number of important factors remain undecided. These affect the way in which the ingredient declaration is to be made. The EC Reference List of Ingredients (i.e. the list of main ingredients normally used in manufacture of compound feedingstuffs) and the provision on the use of abbreviated terms have yet to be finalised and agreed. Thus, those compounders who wish to declare separate ingredients may use their own descriptions until such time as the official reference list is adopted. However, compounders may wish to minimise the changes required when the Reference List is agreed and introduced later next year by listing ingredients in accordance with the latest draft.

Categories Listing

Those compounders who wish to declare ingredients by categories must use the names as listed in Schedule 6 of the Regulations. If one of the ingredients belongs to none of the categories, then that ingredient, designated by its specific name shall be listed in order by weight in relation to the categories. The use of abbreviated terms for categories is not permitted.

Inclusion of Additives and Certain Protein Sources in the List of Ingredients

The use and declaration of both additives and certain protein sources is covered by specific schedules in the Regulations. Thus, it is not necessary to list these products in terms of either categories or individual ingredients, since there are specific declarations and labelling requirements already in force. The only mineral inclusions that have to be included in the "Declaration of Ingredients" are the macro ingredients such as "calcium carbonate", "magnesium oxide" and "mono- dicalcium phosphate".

Enforcement of the New Regulations

The Declaration of Ingredients

New powers have been introduced into the Feedingstuffs Regulations 1991 which extend the existing powers under Section 76 of the Agriculture Act 1970. Consequently, not only will the local authority inspectors be empowered to enter premises under circumstances related to the sale and purchase of feedingstuffs and take samples for analysis, they will also be able to inspect the manufacturers written records. These new powers will, for example, enable inspectors to verify ingredient listing. They will also facilitate checks to ensure safety provisions are being observed. These include both the specific requirement on undesirable substances as well as the general provision that feedingstuffs should not be unwholesome or a danger to animal and human health.

The local authority inspectors will be able to:

(a) Require the compound feed manufacturer to produce any record, in written or any other form, relating to the manufacture by that person of any such compound feedingstuff;

(b) Inspect and take copies of any such record.

The Introduction of a Transitional Period

Discussions have been held between MAFF, LACOTS and UKASTA during which the Association requested LACOTS to consider asking Chief Officers to exercise their discretion on full enforcement of the Feeding Stuffs Regulations 1991 for a period of six months from 22 January 1992.

LACOTS have, therefore, asked Enforcement Authorities to exercise their discretion on this matter to permit in practice a six month Transitional Period to operate. MAFF have indicated that they would not object to this approach. LACOTS have reminded UKASTA that this approach would be subject to the caveats that member companies take all steps to comply with the new legislation as soon as possible and that companies anticipating major problems contact their "home authority" immediately.

Optional Declarations

Most of the optional declarations remain unchanged. The one exception is that for the energy content of poultry compound feedingstuffs which is now to be optional.

Declarations of energy on cattle/sheep and pig compound feedingstuffs will be optional provided that a nationally recognised system is used, such as E3 for ruminants.

The three new equations have now been agreed as follows:

(a) Poultry Feeds [megajoules (MJ) of metabolisable energy (ME) nitrogen corrected per kilogram of compound feed]:

$$\begin{aligned} MJ\ ME/kg\ feed &= 0.1551 \times \%\ Crude\ Protein \\ &+ 0.3431 \times \%\ Oil\ (OAH) + 0.1669 \times \%\ Starch \\ &+ 0.1301 \times \%\ total\ sugar\ (expressed\ as\ sucrose) \end{aligned} \tag{11.1}$$

(b) Ruminant Feeds [megajoules (MJ) of metabolisable energy (ME) per kilogram of dry matter in the compound feed]:

$$\begin{aligned} MJ\ ME/kg\ DM &= 0.14 \times \%\ Neutral\ detergent\ Cellulase \\ &plus\ Gamanase\ and\ Amylase Digestibility \\ &+ 0.25 \times \%\ Oil\ (OAH) \end{aligned} \tag{11.2}$$

(c) Pig Feeds [megajoules (MJ) of digestible energy (DE) per kilogram of dry matter in the compound feed]:

$$MJ\ DE/kg\ DM = 17.47 + 0.079 \times \%\ Crude\ Protein + 0.158 \times \%\ Oil\ (OAH) - 0.331 \times \%\ Ash - 0.140 \times \%\ Neutral\ Detergent\ plus\ Amylase\ Fibre \quad (11.3)$$

All three equations will become optional from 22 January 1992 as far as UK legislation is concerned and it is important that the same equations are used throughout the industry and especially by consultants who monitor energy levels in feeds.

Several problems have been encountered including the variability of OAH (acid hydrolysis of oil) results from some laboratories, and variabilities arising from quality standards of the enzymes used for digestibility measurements.

Continuous collaborative testing is one way of monitoring performance to check on bias and variance of results.

Dietary feedingstuffs

During EEC discussions concerning the directive (90/44) amending the Marketing of Compound Feeds Directive 79/373, the following report was recorded at a session of the Council of Ministers in January 1990:

"The Council is aware of the importance in the market of compound feedingstuffs intended to meet the special nutritional needs of certain categories of animal and would request the Commission to submit proposals to it as quickly as possible, designed to supplement existing Community rules with provisions laying down the specific rules to be observed during the production, marketing and checking of compound feedingstuffs for animals with particular nutritional needs, more commonly known as dietetic animal feedingstuffs".

This new classification of feeds is directed at two main categories, namely feeds for the maintenance of or feeds for the restoration of the physiological status of the animal.

Generally speaking, the use of the feedingstuffs in question is for a limited period only, because they are to treat a temporarily modified physiological state. However, an exception to this rule is provided for feedingstuffs intended for animals whose process of assimilation or metabolism is "irreversibly" affected (diabetic animals or animals suffering from chronic kidney problems). In these cases, it is recommended in the instructions for use of the feedingstuff to seek the advice of a veterinarian before administering the feedingstuff to the animal for a prolonged period.

The draft is in the form of a regulation but some member states, including the UK, favour directive status for this new legislation. The key definitions currently under discussion are as follows, but will be subject to further consideration:

(a) 'feedingstuffs for particular nutritional purposes' means feedingstuffs which, by virtue of their particular composition or method of manufacture are intended for specific nutritional needs. They differ from the products covered by Directive 79/373/EEC of 2 April 1979 on the marketing of compound feedingstuffs and from those defined in Directive 90/167/EEC of 26 March 1990 laying down the conditions governing the preparation, placing on the market and use of medicated feedingstuffs in the Community.

(b) 'Particular nutritional purpose' means to satisfy the maintenance or restoration of the physiological balance of domestic pets or productive livestock.

(c) 'Maintenance or restoration of physiological balance' means exerting a favourable effect on:

1. Certain categories of animals whose assimilation or metabolism is impaired;
2. Certain categories of animals temporarily in particular physiological conditions and which are therefore able to derive benefit from controlled ingestion of feedingstuffs appropriate to their condition;
3. Animals suffering from problems associated with ageing;
4. Animals whose energy requirements are particularly high because of the specific functions they perform.

There are detailed proposals for labelling which are in line with the current requirements for feeds with additional information on the "conditions" being "treated", and the special nutritional characteristics of the feed.

UKASTA had already submitted a number of points to MAFF listing the issues that should be resolved, such as the criteria to distinguish between dietary and medicated feeds, registration of manufacturers, training and qualifications of those who formulate and sell dietary feeds etc. The creation of this new category is necessary in part to permit compounders to manufacture feeds with levels of trace elements, vitamins and other ingredients which are prohibited under current legislation by absolute maxima in, for instance, the additives directives, e.g. trace element and vitamin maxima which cannot be exceeded even under a VWD because the products are not licensed as medicinal products and cannot therefore be prescribed.

The legislation is likely to be finalised during 1992.

The Food Safety Act 1990

The Food Safety Act 1990 came into force on 1 January 1991. It is a major new piece of legislation aimed at protecting the consumer. It covers the whole food chain and therefore farmers and growers will be directly affected.

Anybody who sells food for human consumption is covered by the Food Safety Act. The definition of food under the Act includes such things as crops, fruit and vegetables as soon as they are harvested, eggs as soon as they are collected and raw milk. Many farmers and growers are therefore included. They may be selling direct to the consumer from a farm shop, or to processors, wholesales, retailers, other farmers or anybody else.

It is an offence to render food injurious to health. It is also an offence to sell, or keep for eventual sale, food that:

(a) Is unfit for human consumption;

(b) Has been rendered injurious to health;

(c) Is so contaminated that it would be unreasonable to expect people to eat it in that state;

(d) Is not of the nature, substance or quality demanded by the purchaser;

(e) Is falsely or misleadingly labelled or presented.

These offences apply only to food for human consumption.

The Food Safety Act does not only cover food. It also enables Ministers to make regulations covering "food sources". These are growing crops or live animals from which it is intended to derive food. Ministers intend to use these powers to increase controls over the use of veterinary medicines in animals. Members of the State Veterinary Service would, under the proposed regulations, be able to take action on farms where they found that illegal substances had been used or maximum permitted levels of residues of medicinal or other substances in meat had been exceeded at the point of slaughter.

The legislation is already drafted concerning maximum residue limits of medicinal products in animal products for human consumption.

During 1991 two major sets of new regulations were being prepared. Under the first, all premises used by food businesses will have to register with their local authority. These will mainly be manufacturing, retailing and catering premises and some farms will be included. The exact details and coverage will be decided after consultation. Registration will be straightforward. There will be a simple form to complete and there will be no charge.

The second set of new regulations will require people who handle food commercially to be trained to an appropriate level in food hygiene. The aim is to ensure that people have the practical skills and knowledge they need to keep food safe. This requirement is also directed mainly at manufacturers, retailers and caterers but some farmers and growers will be included. Training requirements will be set at appropriate levels after consultation.

The new Act has very important implications for feed compounders. Although animal feedingstuffs are not covered by the Act, they can directly affect the safety and quality of "food sources", namely animals intended for human consumption. It is therefore incumbent upon feed manufacturers to ensure that the feed is of the required quality and safety with relevant warnings on labels where appropriate and sound advice of feeding programmes. The main areas are the avoidance of crosscontamination with medicinal products which could lead to high residues in animal products, the availability of "withdrawal" feeds, and proof of adequate quality assurance. This latter consideration is very important because a legal defence of "due diligence" is available under the Act. In the event that prosecutions were pursued, documentary evidence that appropriate quality assurance systems were effectively applied as a routine would be a powerful and admissible defence. BS5750 registration of a feed compounders operations covering all aspects from raw material purchasing to delivery of the feed is regarded as particularly good evidence of effective systems being operated.

The new law relating to food will mostly be enforced by local authority officers. The Government will issue a code of practice to guide local authorities on how food safety law in relation to farms is to be enforced. This code will clarify when it might be appropriate for officers to go on farms and how inspections should be conducted. Farming and other organisations will be consulted before the code is issued.

Local authority enforcement officers will have tougher powers to seize and detain suspect food and will be able to issue formal notices requiring improvements to premises, equipment and practices which breach regulations. If necessary, they will be able to require the closure of operations which pose an imminent danger to health. These powers apply only to the new food law. They do not change the operation of existing legislation, for example that covering animal health or pesticides. The Food Safety Act contains a right of appeal against the actions of enforcement officers. If a court decides that officers have used their powers wrongly, compensation will be paid.

Ministers have new powers to make emergency control orders where there is an imminent risk of injury to health.

The Environmental Protection Act

The major features of the legislation are:

- *Part B Process* - The manufacture of animal feedingstuffs is a Part B process and thus compounders would have to control their emissions of dust and odour to the atmosphere.

- *Timetable* - Applications for authorisation of existing Part B processes will have to be made between 1 April 1992 and 30 September 1992 for England and Wales, and between 1 December 1992 and 31 March 1993 for Scotland.

- *Charges* - The charge for Part B processes, which are subject to local authority control, will be:

  (a) Application fee: £800

  (b) Annual (or "subsistence") charge : £500 - payable (pro-rata) on receipt of authorisation and thereafter on 1 April each year.

  (c) Subsequent variations : £530.

Guidance Note - In order to enable feed compounders to meet the requirements of the legislation, as well as ensure consistent enforcement through the country, the Government will publish through HMSO a document entitled "Secretary of State's Guidance - Animal Feed Compounding Processes".

The UKASTA Legislation Sub-Committee, together with representatives from the Engineering and Production Management sector of member companies were involved in negotiations, during the first part of 1991, with representatives of Her Majesty's Inspectorate of Pollution and the local authorities (Environmental Health Department) on the draft Guidance Note. The Association successfully negotiated significant modifications to the draft Guidance Note. They were able to demonstrate that the total particulate matter in all emissions to air from product coolers should be 150 mg/m^3 and not 50 mg/m^3 as originally proposed. (The original level of 50 mg/m^3 still applies to emissions from all other sources.) Also, when monitoring, sampling and measuring emissions, it was agreed that in the case of the exhaust vents which emitted low volumes of air flows within the environs of the building, these should be subject to regular manual inspection. (It was originally proposed that continuous monitoring equipment should be installed on such exhaust vents.) Exhaust vents on grinders and coolers will require continuous monitoring.

The final draft Guidance Note has now been issued by the DoE. Some of the key clauses in the Guidance Note are shown in Appendix B.

Undesirable substances: EEC proposals

A Commission proposal was published on 6 November 1991 which will amend Directive 74/63/EEC on Undesirable Substances in Animal Feed.

This Directive sets maximum permitted levels for a range of substances (heavy metals, organo-chlorine pesticides, mycotoxins) which may adventitiously contaminate straight and compound feedingstuffs. In addition to levels in finished feeds there are also maximum levels set for aflatoxin B1 in six raw materials and cadmium, arsenic, lead and fluorine in phosphate raw materials.

The Commission proposals would:

(a) Extend the control of undesirable substances in certain raw materials by introducing further maximum levels (for Arsenic, Lead, Fluorine) and reducing current levels for aflatoxin and cadmium.

(b) Prohibit the mixing of consignments of raw materials when one or more is above the maximum permitted level (i.e. restrict blending down to achieve a legal level).

(c) Provide for consignments of raw materials to be labelled with the level of undesirable substance, where this level is higher than the level for finished feed. These consignments may only be traded through compounders on a national register.

(d) Tighten up provisions for information exchange between enforcement authorities about highly contaminated consignments.

(e) Extend the scope of the Directive to wild animals.

The proposal is in response to the lead in feed incident in 1989 and a commitment in Directive 74/63 to remove national controls which create trade barriers. The new directive must be implemented by 31 December 1993.

The proposals relating to maximum levels on undesirables in raw materials are not currently very extensive. However the quantitative labelling of all contaminants in raw materials will have a considerable impact on importers and merchants. The strengthening of controls may also have trading repercussions for third country suppliers.

Miscellaneous legislation

Prohibited ingredients in animal feeds - The list of ingredients whose use is prohibited in animal feedingstuffs was published on 9 October 1991 (91/516/EEC) and are:

(a) Faeces, urine as well as separated digestive tract content resulting from the emptying or removal of the digestive tract, irrespective of any form of treatment or admixture.

(b) Leather and leather waste.

(c) Seeds and other plant propagating materials which, after harvest, have undergone specific treatment with plant protection products for their intended use (propagation), and any derived by-products.

(d) Wood, sawdust and other materials derived from wood treated with wood protection products.

(e) Sludge from sewage plants treating waste waters.

Aflatoxin B1 - MAFF have confirmed that there is to be a short delay in the introduction of the revised levels of aflatoxin B1 in complementary dairy feedingstuffs and certain straight feedingstuffs. The legislation was to have been brought into operation from 30 November 1991. It is, however, being included in the consolidated Feedingstuffs Regulations 1991 which are not yet available. Thus, MAFF advise that the new aflatoxin controls will be coming into force on 22 January 1992.

A summary of the changes to be introduced are as follows:

(a) The reduction of the current maximum limit set for the following straight feedingstuffs, groundnut, copra, palm kernel, babussu, cottonseed, and maize, from 0.05 mg/kg to 0.02 mg/kg.

(b) The reduction of the current maximum limit set for other complementary feeds, i.e. dairy feeds from 0.01 mg/kg to 0.005 mg/kg.

Pathogens - The EEC Pathogens Directive is making slow progress. FEFAC have amended the UK Code for control of salmonella in feeds and submitted it to the Commission as a key part in the proposed controls. The directive should be finalised by December 1992

Conclusion

New and amended legislation covering all aspects of feed compounding continues to exercise a primary influence and cost on the industry. Further market pressures to demonstrate effective quality systems are also increasingly linked to high profile legislation. This trend is likely to continue and encompass more compounders and livestock producers.

Appendix A

New style VWD for medicated feeds

DIRECTION FOR THE INCORPORATION OF A MEDICINAL PRODUCT IN AN ANIMAL FEEDINGSTUFF OR FOR THE PLACING ON THE MARKET OR IMPORTATION OF MEDICATED ANIMAL FEEDING STUFFS VETERINARY WRITTEN DIRECTION

This Direction shall not be re-used.

Reference No.

SECTION I - TO BE COMPLETED IN ITS ENTIRETY BY VETERINARY SURGEON OR VETERINARY PRACTITIONER

1. Please manufacture/sell/supply/import* .... tonnes/kg* of (name/type of feed)

.....................meal/pellets/crumbs* containing -

..............) ....) (proprietary name(s) and

..............) g/tonne (mg/kg)* of ....) product licence number(s)

..............) ....) and/or generic name(s)

to give in total -

.............) ....) precise/description of

.............) g/tonne (mg/kg)* of ....) active substance(s))

in the final medicated feeding stuff for administration to the following animals which are under my care:

Species......................... Approx. number............................

Note. The amount shall not exceed 31 days' supply.

2. The medicated feeding stuff must be sold/supplied* to (name of farmer and address of farm)

....

....

Recommendations for Use On the Farm

(i) Disease to be treated...

(ii) Quantity of medicated feeding stuff to be given daily................

....

(iii) Duration of treatment...

(iv) Animals must not be slaughtered for human consumption until..........

.... after the last treatment.

Milk/eggs* must not be taken for human consumption until............

.... after the last treatment.

(v) Special precautions...

3. This direction is valid for 31 days from the date of signature.

Signature of Veterinary Surgeon

or Veterinary Practitioner ....

Name in block letters ....

Practice Address ....

Date.................Tel No ....

SECTION II - TO BE COMPLETED BY VETERINARY SURGEON OR PRACTITIONER OR FARMER

Name and address of manufacturer/seller supplier/importer*

....

....

....

SECTION III - TO BE COMPLETED BY THE MANUFACTURER/SELLER/SUPPLIER

Date(s) of delivery...

To be used before...

Signature of manufacturer/seller/supplier*

....

SECTION IV - IF APPLICABLE, TO BE COMPLETED BY VETERINARY SURGEON OR VETERINARY PRACTITIONER

1. Reason(s) for authorising incorporation by a manufacturer (including an on-farm mixer) not in the appropriate Part of the Register

....

....

....

*Delete as appropriate.

2. Reason(s) for authorising incorporation by a manufacturer of an unlicensed combination of medicinal products........................

....

....

NOTES

1. This form must be completed in triplicate, in ink or by other indelible means, and signed in ink in his own name by the Veterinary Surgeon or Veterinary Practitioner, who will retain one copy and give one copy each to the manufacturer and the farmer.

2. If any part of Section IV has been completed, the manufacturer must send a copy of the form to the Royal Pharmaceutical Society of Great Britain, 1 Lambeth High Street, London SE1 7JN, or the Department of Agriculture, Northern Ireland, "Duniris", 15 Galway Park, Dundonald, Belfast BT16 0AN, within 28 days of incorporation.

3. A telefacsimile copy may be sent to the incorporator in an emergency. The original veterinary written direction must be sent within 72 hours.

Appendix B

Environmental Protection Act 1990 - Extracts from Guidance Note

MISSION LIMITS AND CONTROLS

11. All emissions to air, other than steam or water vapour, should be colourless and free from persistent mist. All emissions to air should be free from persistent fume and free from droplets.

12. All emissions should be free from offensive odour outside the process boundary, as perceived by the local authority Inspector.

13. Emissions from combustion processes in normal operation should be free from visible smoke and in any case should not exceed the equivalent of Ringelmann Shade 1 as described in British Standard BS 2742:1969.

15. The following concentrations of total particulate matter in all emissions to air should not be exceeded:

From product coolers	150 mg/m^3
From all other sources	50 mg/m^3

16. The introduction of dilution air to achieve emission concentration limits is not permitted, Exhaust flow rates should be consistent with efficient capture of emissions, good operating practice and meeting the requirements of the legislation relating to the workplace environment.

MONITORING, SAMPLING AND MEASUREMENT OF EMISSIONS

17. As part of proper supervision the operator should monitor emissions and make tests and inspections of the process. The need for and scope of testing and the frequency and time of sampling, will depend on local circumstances, operational practice, and the scale of operation. The following should normally be included as standard requirements.

18. Subject to the requirements of Clause 19 emissions from arrestment plant where the exhaust flow rate exceeds $50m^3$ per minutes should be continuously monitored for particulate matter, for example using an optical instrument and the results continuously recorded. Where this arrestment plant includes a bag filter, it should be acceptable to continuously monitor the performance of the bag filter using a bag failure indicator. In this case, the bag failure monitor should be fitted with a visual and audible alarm. This requirement does not apply to emissions from wet scrubbers.

19. In the case of emissions from arrestment plant (other than grinder exhausts) where the exhaust flow rate is less than $100m^3$/min and the exhaust from the arrestment plant discharges into the workplace, the requirements of Clauses 18 and 21 should not apply. In these circumstances arrestment equipment should be inspected regularly and at least once a week and serviced at least once every three months.

20. All continuous monitoring instruments should be checked daily and calibrated in accordance with the manufacturers recommendations, and at least once a year.

21. Subject to the requirements of Clause 19 emissions should be tested at least once a year for particulate matter. The reference test method for particulate matter emissions in chimneys or ducts is that of British Standard BS 3405:1983, and tests should be carried out according to the main procedural requirements of that standard. The results of these tests should be forwarded to the local authority within 8 weeks of the completion of the sampling.

MATERIALS HANDLING

27. All outdoor bulk storage of dusty, or potentially dusty, materials should be stored in silos. The silos should be fitted with arrestment plant, for example a bag filter, to meet the requirements of Clause 15. The storage of raw materials should be permitted inside processing buildings provided that adequate steps are taken to prevent entrainment of particulate matter outside the building, for example by the use of plastic strip curtains on building access points.

28. Adequate provisions should be made for the containment of liquid and solid spillages. All spillages should be cleared as soon as possible and in the case of solid materials this should be achieved by the use of vacuum cleaning, wet methods, or other appropriate techniques. Dry sweeping of dusty spillages should not be permitted.

33 Where raw materials are delivered to bulk storage by tipper lorries, the raw material acceptance area should be provided with protection against wind shipping of particulate matter (ie a covered, screened area) and local exhaust ventilation should be installed and emissions should be discharged through suitable arrestment plant, for example a bag filter, to meet the requirements of Clause 15.

34 The packing of potentially dusty or dry animal feed into bags should be carried out using purpose designed plant, fitted with extraction, discharging through suitable arrestment plant, for example bag filters, to meet the requirements of Clause 15.

35. Where materials are removed from site in bulk vehicles, the loading area should be provided with protection against wind entrainment of particulate matter, for example carried out in a covered, screened area. The discharge of products into the vehicles should be carried out in order to minimise the generation of airborne dust, and fall heights from discharge pipes should be reduced for example by the use of flexible hoses or socks. These areas should be fitted where necessary with local exhaust ventilation discharging through suitable arrestment plant, for example bag filters, to meet the requirements of Clauses 11, 12 and 15.

LIST OF PARTICIPANTS

The twenty-sixth Feed Manufacturers Conference was organized by the following committee:

Dr L.G. Chubb	(Private Consultant)
Mr J. Fordyce	(West Midland Farmers)
Dr K. Jacques	(Alltech Inc)
Mr P. Marriage	(W. & H. Marriage & Sons Ltd)
Dr S. Marsden	(Dalgety Agriculture Ltd)
Mr R.T. Pass	(Thomas Borthwick & Co Ltd)
Mr F.G. Perry	(BP Nutrition (UK) Ltd)
Mr J.R. Pickford	(Tecracon Ltd)
Mr M.H. Stranks	(MAFF, Bristol)
Dr B. Vernon	(Pauls Agriculture)
Dr D.R. Williams	(BOCM Silcock Ltd)

Dr K.N. Boorman
Prof P.J. Buttery
Dr D.J.A. Cole (Chairman)
P.C. Garnsworthy (Secretary)
Dr W. Haresign
Prof G.E. Lamming
Dr J. Wiseman
} University of Nottingham

The conference was held at the University of Nottingham School of Agriculture, Sutton Bonington, 5th-7th January 1992 and the committee would like to thank all the authors for their valuable contributions. The following persons registered for the meeting:

Abma, Ir J.	Orfam BV, Peppelkade 56, 3992 AK, Houten, Holland
Adams, Dr C.A.	Kemin Europa NV, Industriezone, Wolfstee, 2200 Herentals, Belgium
Alderman, Mr G.	University of Reading, Dept. of Agriculture, Earley Gate, Whiteknights, Reading RG6 2AT
Allder, Mr M.J.	Eurotec Nutrition Ltd, Glendale House, 5b Marins Lane, Witchem, Ely, Cambridge CB6 2LB
Allen, Dr J.D.	Frank Wright Ltd, Blenheim House, Blenheim Road, Ashbourne, Derbyshire DE6 1HA
Angold, Mr M.	Roche Products Ltd, Delves Road, Heanor, Derbyshire DE7 7SG
Ashley, Dr J.H.	Rhone Poulenc Animal Nutrition, 20 Av. Raymond Aron, 92165 Antony Cedex, France
Aspland, Mr F.P.	Aspland & James Ltd, 118 Bridge Street, Chatteris, Cambs PE16 6RN
Azzi, Miss D.	Sopropeche, B.P.275, 62204 Boulogne sur Mer Cedex, France

Barrie, Mr M.J.	Elanco Animal Health, Dextra Court, Chapel Hill, Basingstoke, Hants RG21 2SY
Bates, Mrs M.	Vitrition Ltd, Ryhall Road, Stamford, Lincs
Bedford, Dr M.	Finnfeeds International Ltd, Forum House, 41-51 Brighton Road, Redhill, Surrey
Beer, Mr J.H.	W & J Pye Ltd, Fleet Square, Lancaster LA1 1HA
Beer, Dr J.V.	The Game Conservancy, Fordingbridge, Hants SP6 2RA
Belyavin, Dr C.	Chris Belyavin (Technical) Ltd, 2 Pinewoods Church, Aston, Newport, Shropshire TF10 9LN
Bentley, Dr J.S.	British United Turkeys, Hockenhull Hall, Tarvin, Chester CH3 8LE
Berry, Mr M.H.	Berry Feed Ingredients Ltd, Chelmer Mills, New Street, Chelmsford, Essex CM1 1PN
Best, Mr P.	Feed International, 18 Chapel Street, Petersfield, Hampshire GU32 3DZ
Blanch, Mr A.	Nutricio i Alimentacio Animal, Facultat Veterinaria-VAB, 08193-Bellaterra (Barcelona), Spain
Bliss, Mrs. H.J.	Carrs Agriculture Ltd, Old Croft, Stanwix, Carlisle
Bole, Mr J.	David Patton Ltd, Milltown Mills, Monaghan, Ireland
Boorman, Dr K.N.	University of Nottingham, School of Agriculture, Sutton Bonington, Loughborough, Leics LE12 5RD
Boothroyd, Miss J.	Frank Wright Ltd, Blenheim House, Blenheim Road, Ashbourne, Derbys DE6 1HA
Bouchard, Mr K.A.	Midland Shire Farmers Ltd, Defford Mill, Earls Croome, Worcester WR8 9DF
Bourne, Mr S.	Alltech S.E. Asia, c/o Biotechnology Centre, 3031 Catnip Hill Pike, Nicholasville, KY 40356, USA
Boyd, Dr P.A.	Cranswick Mill Ltd, The Airfield, Cranswick, Driffield, N Humberside
Brenninkmeijer, Dr C.	Hendrix' Voeders BV, PO Box 1, 5830 MA Boxmeer, The Netherlands
Britten, Miss S.	Sun Valley Poultry Ltd, Sun Valley Feed Mill, Tram Inn, Allensmore, Hereford HR2 9AW
Brooking, Miss P	W J Oldacre Ltd, Cleeve Hall, Bishops Cleeve, Cheltenham, Glos
Brophy, Mr A.	Alltech Ireland, Unit 28, Cookstown Industrial Estate, Tallaght, Co. Dublin
Brown, Mr G.	Colborn Dawes Nutrition Ltd, Heanorgate, Heanor, Derbys DE7 7SG
Browne, Dr J.W.	ICI Nutrition, Alexander House, Runcorn, WA7 2UP
Bruce, Dr D.W.	Devenish Feed Supplements Ltd, 96 Duncrue Street, Belfast BT3 9AR
Brumby, Dr P.E.	Amalgamated Farmers Ltd, Kinross, New Hall Lane, Preston PR1 5JX
Burley, Mrs J.K.	Chapman & Frearson Ltd, Victoria Street, Grimsby, S. Humberside DN31 1PX
Burt, Dr A.W.A.	Burt Research Ltd, 23 Stow Road, Kimbolton, Huntingdon, Cambs PE18 0HU
Bush, Mr T.	International Additives Ltd, Old Gorsey Lane, Wallasey, Merseyside L44 4AH

Buttery, Prof P.J.	University of Nottingham, School of Agriculture, Sutton Bonington, Loughborough, Leics LE12 5RD
Buysing Damste, Ir B.	Trouw International BV, Research & Development, P O Box 50, 3880 AB Putten, Holland
Campani, Dr I.	F.LLI Martini e C. s.p.a., 2614 Emilia Street, Budrio di Longiano-Fo, Italy
Carmichael, Mr A.	Elanco Animal Health, Dextra Court, Chapel Hill, Basingstoke, Hants RG21 2SY
Carruthers, Miss S.A.	Pauls Agriculture Ltd, Lindum Mill, Shepshed, Leics.
Carter, Dr T.J.	Kemin UK Ltd, Becor House, Green Lane, Lincoln LN6 9DC
Caygill, Dr J.	MAFF Chief Scientists Group, Room G9, Nobel House, 17 Smith Square, London SW1P 3JR
Charles, Dr D.R.	M A F F, Block 7, Chalfont Drive, Nottingham NG8 3SN
Charlton, Mr P.	Alltech UK, 16/17 Abenbury Way, Wrexham Industrial Estate, Wrexham, Clwyd LL13 9UZ
Chaudhry, Dr A.S.	University of Bristol, Veterinary School, Kangford, BS18 7DU
Choi, Mr B.S.	Taihan Sugar Industrial Co Ltd, South Korea
Chubb, Dr L.G.	Koonunga, 39 Station, Road, Harston, Cambridge CB2 5PP
Clarke, Mr A.N.	Four-F Nutrition, Darlington Road, Northallerton, N.Yorks DL6 2NW
Clay, Mr J.	Alltech UK, Unit 16/17 Abenbury Way, Wrexham Industrial Estate, Wrexham, Clwyd LL13 9UZ
Close, Dr W.H.	Close Consultancy, 129 Barkham Road, Wokingham, Berks RG11 2RS
Coe, Mr P.C.	Smithkline Beecham Animal Health Ltd, Hunters Chase, Walton Oaks, Tadworth, Surrey KT20 7NT
Cole, Dr D.J.A.	University of Nottingham, School of Agriculture, Sutton Bonington, Loughborough, Leics LE12 5RD
Cole, Mr J.R.	International Additives Ltd, Old Gorsey Lane, Wallasey, Merseyside L44 4AH
Cole, Mr M.A.	BOCM Silcock, Basing View, Basingstoke, Hampshire RG21 2EQ
Colenso, Mr J.	BP Nutrition (UK) Ltd, Wincham, Northwich, Cheshire CW9 6DF
Connolly, Mr J.G.	Wm Connolly & Sons Ltd, Red Mills, Goresbridge, Co. Kilkenny, Ireland
Cooke, Dr B.	Dalgety Agriculture Ltd, 180 Aztec West, Almondsbury, Bristol BS12 4TH
Cooper, Dr A.	Seale Hayne Polytechnic South West, Newton Abbot, Devon TQ12 6NQ
Cooper, Mr R.	Finnfeeds International Ltd, Forum House, 41-51 Brighton Road, Redhill, Surrey
Cooper, Miss S.E.	Butterworth Heinemann Ltd, Linacre House, Jordan Hill, Oxford OX2 8DP
Courtin, Mr B.	EMC Belgium, Square de Meeus 1, 1040 Brussels, Belgium
Cowan, Dr D.	Novo Nordick (UK) Ltd, 4 St Georges Yard, Castle Street, Farnham GU9 7LW
Cox, Mr C.N.	SC Associates (Feedingstuffs) Ltd, The Limes, Sowerby Road, Sowerby, Thirsk, N Yorks YO7 1HX
Cullin, Mr A.W.R.	Forum Chemicals Ltd, Forum House, 41-52 Brighton Road,

	Redhill, Surrey
Dakin, Mr R.G.	Novus UK Ltd, Cotteswold House, 14 Gloucester Street, Cirencester, Glos GL7 2DG
Dann, Mr R.	Feed Flavours (Europe) Ltd, Waterlip Mill, Cranmore, Shepton Mallet, Somerset
Davies, Mr J.	North Eastern Farmers Ltd, Bannermill, Aberdeen AB9 2QT
Davies, Dr J.L.	Colborn Dawes Nutrition, Heanorgate, Heanor, Derby DE7 7SG
Dawson, Mr W.	Britphos Ltd, Rawdon House, Green Lane, Yeadon, Leeds LS19 7BY
De Bruyne, Mr K.	EMC Belgium, Square de Meeus 1, 1040 Brussels, Belgium
De Vriendt, Dr D.L.	Radar, Dorpsstraat 4, 9800 Deinze, Belgium
Dean, Mr J.M.	AGIL Ltd, Gercules 2, Calleva Park, Aldermaston
Dean, Mr R.W.	Dean Agricultural Associates, 16 St Georges Lodge, London N10 3TE
Diepenbroek, Mr L.D.	Mole Valley Farmers, Station Road, South Molton, Devon
Dixon, Mr D.H.	Brown & Gillmer Ltd, Seville Mills, Seville Place, Dublin 1
Douglass, Mrs V.L.	Bernard Matthews plc, Gt Witchingham Hall, Norwich, Norfolk NR9 5QD
Dunne, Mr S.B.J.	Fishers Nutrition Ltd, Driffield, East Yorkshire YO25 9PF
Edwards, Mr A.A.	Elanco Animal Health, Dextra Court, Chapel Hill, Basingstoke, Hants EG21 2SY
Ewing, Dr W.	International Molasses, Division of Cargill plc, King George Dock, Hull HU9 5PR
Fairey, Mr M.B.	Dalgety Agriculture Ltd, 180 Aztec West, Bristol BS12 4TH
Farley, Mr R.L.	BP Nutrition (UK) Ltd, Wincham, Northwich, Cheshire
Farrell, Dr D.J.	University of New England, Department of Biochemistry, Microbiology & Nutrition, Armidale, 2351 NSW, Australia
Fawcett, Dr R.	Edinburgh University, School of Agriculture, Kings Buildings, West Mains Road, Edinburgh EH9 3JG
Fawcett, Mr T.J.	Amalgamated Farmers Ltd, Kinross, Newhall Lane, Preston, PR1 5JX
Felisaz, Mr D.	Laboratories Pancosma SA, Voie des Traz 5-6, CH-1218 Grand-Saconnex/GE
Filmer, Mr D.	David Filmer Ltd, Wascelyn, Brent Knoll, Somerset TA9 4DT
Fisher, Dr C.	Leyden Old House, Kirknewton, Midlothian EH27 8DQ
Fitt, T D.R.	Colborn Dawes Nutrition Ltd, Heanorgate, Heanor, Derbys DE7 7SG
Fletcher, Mr C.J.	Aynsome Laboratories, Kentsford Road, Grange-over-Sands, Cumbria
Fordyce, Dr J.	West Midland Farmers, Bradford Road, Milksham, Wiltshire SN12 8LQ
Fox, Mr P.A.	David Moore (Flavours)Ltd, 29 High Street, Harpenden, Herts AL5 2RU
Fraser, Mrs S.	International Milling (Flour & Feed), Turret House, 171 High Street, Ricksmansworth, Herts WD3 1SN
Fullarton, Mr P.J.	Forum Chemicals Ltd, Forum House, Brighton Rd, Redhill, Surrey RH1 67S
Gaisford, Mr M.	Farmers Weekly, Quadrant House, The Quadrant, Sutton,

	Surrey
Garnsworthy, Dr P.C.	University of Nottingham, School of Agriculture, Sutton Bonington, Loughborough, Leics. LE12 5RD
Geary, Mr B.	Hoechst UK Ltd, Hoechst Animal Health, Walton Manor, Walton, Milton Keynes, Bucks MK7 7AJ
Geddes, Mr N.	Nutec Limited, Eastern Avenue, Lichfield, Staffs
Geerse, Ir C	Royal Gist Brocades NV, Agro Business Group, PO Box 1820, 2280 DV Rijswijk, The Netherlands
Gibson, Mr J.E.	Parnutt Foods Ltd, Hadley Road, Woodbridge Industrial Estate, Sleaford, Lincs. NG34 7EG
Gilbert, Mr R.	Asbury Publications Ltd, Stoke Road, Bishop's Cleeve, Glos. GL52 4RW
Gillespie, Miss F.T.	United Molasses, Stretton House, Derby Road, Stretton, Burton on Trent, Staffs DE13 0DW
Gjefsen, Dr T.	nNorske Felleskjop, Lille Grensen 7, 0159 Oslo I, Norway
Gooderham, Mr B.J.	Pye Milk Products, W & J Pye Ltd, Lansil Industrial Estate, Caton Road, Lancaster LA1 3PQ
Gould, Mrs M.P.	Volac Ltd, Orwell, Royston, Herts
Grace, Mr J.	Elanco Animal Health, Dextra Court, Chapel Hill, Basingstoke, Hants RG21 2SY
Graham, Dr H.	Finnfeeds International Ltd, 41-51 Brighton Road, Redhill, Surrey RH1 6YS
Gray, Mr W.	Kemira Kemi (UK) Ltd, Orm House, Hookstone Park, Harrogate, North Yorks
Green, Dr S.	Rhone Poulenc Chemicals, Oak House, Reeds Crescent, Watford, WD1 1QH
Griffiths, Mr W.D.E.	Midland Shire Farmers Ltd, Defford Mill, Earls Croome, Nr Worcester
Guenther, Dr C.	BASF AG, MEV/LT - D 205, 6700 Ludwigshafen, Germany
Haggar, Mr C.W.	Candus Limited, 1 Park Crescent, Addingham, Ilkley, West Yorkshire LS29 0LX
Hall, Dr G.R.	Kemin UK Ltd, Becor House, Green Lane, Lincoln LN6 9DL
Hannagan, Mr M.J.	West Coates, 11 Durbin Park Road, Clevedon, Avon BS21 7EU
Hardy, Dr B.	Dalgety Agriculture Ltd, 180 Aztec West, Almondsbury, Bristol BS12 4TH
Haresign, Dr W.	University of Nottingham, School of Agriculture, Sutton Bonington, Loughborough, Leics LE12 5RD
Harker, Dr A.J.	Finnfeeds International Ltd, Forum House, 41-51 Brighton Road, Redhull, Surrey
Harrison, Mrs J.S.	Sciantec Analytical Services, Main Site, Dalton, Thirsk, North Yorkshire YO7 3JA
Haythornthwaite, Mr A.	Nu-Wave Health Products Ltd, 45 Church Road, Warton, Preston PR4 1BD
Hegeman, Mr F.	Borculo UK, Bryman Four Estate, River Lane, Saltney, Chester
Hemingway, Prof G.	Glasgow University Veterinary School, Bearsden, Glasgow
Higginbotham, Dr J.D.	United Molasses, Stretton House, Derby Road, Stretton, Burton-on-Trent, Staffs DE13 0DW
Hitchens, Mr C.T.	Favor Parker Ltd, The Hall, Stoke Ferry, Kings Lynn, Norfolk PE33 9SE
Hockey, Mr R.	Smithkline Beecham Animal Health Ltd, Terrapin Building,

	Walton Oaks, Tadworth, Surrey KT20 7HE
Huggett, Miss C.D.	University of Nottingham, School of Agriculture, Sutton Bonington, Loughborough, Leics LE12 5RD
Huisman, Dr J.	IVT-TNO/ILOB, P O Box 15, 6700 AA Wageningen, The Netherlands
Hyslop, Mr J.J.	United Distillers, Biotechnology Group, Carsebridge Road, Alloa, FJ10 3LT
Inborr, Mr J.	Finnfeeds International Ltd, Forum House, 41-51 Brighton Road, Redhill, Surrey
Ingham, Mr R.W.	Kemin UK Ltd, Becor House, Green Lane, Lincoln LN6 9DC
Jacklin, Mr D.	ADAS, Ministry of Agriculture, Fisheries & Food, Kenton Bar, Newcastle-on-Tyne NE1 2YA
Jackson, Mr J.C.	Nutec Limited, Eastern Avenue, Lichfield, Staffs
Jacques, Dr K	Alltech Biotechological Centre, 3031 Catnip Hill Pike, Nicholasville, Kentucky 40356, USA
James, Mr S.	Chief Scientists Group, MAFF, 17 Smith Square, London SW1P 3JR
Janes, Mr R.	BP Nutrition (UK) Ltd, Wincham, Northwich, Cheshire
Jones, Dr E.	Dalgety Agriculture Ltd, 180 Aztec West, Almondsbury, Bristol BS12 4TH
Jones, Mr E.J.	Format International Ltd, Format House, Poole Road, Woking, Surrey
Jones, Miss F.	Cherry Valley Farms Ltd, N Kelsey Moor, Caistor, Lincs. LN7 6HH
Kahrs, Dr D.	Lohmann Tierernahrung GmbH, Postfach 446, Nuefelder Strasse, D-2 190 Cuxhaven 1, Germany
Kennedy, Mr G.	BASF plc, Earl Road, Cheadle Hulme, Cheshire
Ketelaar, Mr G.G.	Pricor B V, Postbus 51, 3420 DB Oudewater, The Netherlands
Keys, Mr J.	32 Holbrook Road, Stratford-upon-Avon, Warwickshire CV37 9DZ
Lamming, Prof G.E.	University of Nottingham, School of Agriculture, Sutton Bonington, Loughborough, Leics LE12 5RD
Lane, Mr P.	Parnutt Foods Ltd, Hadley Road, Woodbridge Industrial Estate, Sleaford NG34 7EG
Law, Mr C.A.	LaFarge Special Products, 730 London Road, Grays, Essex
Law, Mr J.R.	Sheldon Jones Agriculture, Priory Mill, West Street, Wells, Somerset BA5 2HL
Lawrence, Dr K.	Elanco Products Ltd, Dextra Court, Chapel Hill, Basingstoke, Hants RG21 2SY
Lawrie, Mr S.M.	Agricultural & Food Research Council, Polaris House, North Star Avenue, Swindon SN2 1UH
Lee, Mr H.W.	Taihan Sugar Industrial Co Ltd, South Korea
Lima, Mr S.	Felleskjopet Rogaland Agder, P O Box 208, N-4001 Stavanger, Norway
Lowe, Mr J.	Gilbertson & Page, P O Box 321, Welwyn Garden City, Herts, AL7 1AF
Lyons, Dr P.	Alltech Inc., Biotechnology Centre, 3031 Catnip Hill Pike, Nicholasville, Kentucky KY 40356 USA
Mackey, Mr W.S.	G E McLarnon & Sons Ltd, 126 Moneynick Road, Randalstown, Antrim BT41 3HU
MacMahon, Mr M.J.	MacMahon Associates, Clarendon Business Centre, 2/3 Clarendon Terrace, Stockbridge, SO20 6EY
Madsen, Dr J.	Royal Vet. and Agric. University, 23 Rolighedsvej,

	1958 Frederiksberg C, Denmark
Major, Mr N.C.	BOCM Silcock Ltd, The Square, Basing View, Basingstoke, Hants SP10 4ER
Malandra, Dr F.	Sildamin Spa, Sos Tegno di Spessa, 27010 Pavia, Italy
Marangos, Dr A.G.	S J D Humphrey Holdings Ltd, Hazeley Road, Twyford, Winchester, Hants SO21 1QA
Marchment, Dr S.	Format International Ltd, Format House, Poole Road, Woking, Surrey GU21 1DY
Marriage, Mr P.	W & H Marriage & Sons Ltd, Chelmer Mills, Chelmsford CM1 1PN
Marsden, Dr M.	J Bibby Agriculture Ltd, Head Office, Adderbury, Oxon
Marsden, Dr S.	Dalgety Agriculture Ltd, 180 Aztec West, Almondsbury, Bristol BS12 4HQ
Martyn, Mr S.	International Additives Ltd, Old Gorsey Lane, Wallasey, Merseyside L44 4AH
Mather, Mr S.	Feed Flavours (Europe) Ltd, Waterlip Mill, Cranmore, Shepton Mallet, Somerset BA4 4RN
McAllan, Dr A.B.	AFRC Inst. Grassland & Environmental Research, Hurley, Maidenhead, Berks SL6 5LR
McEnroe, Mr P.	Kerry Group plc, Princes Street, Tralee, Co Kerry, Ireland
McGrane, Mr M.	Macroom Mills Ltd, Macroom, Co. Cork, Ireland
McGuckin, Dr E.	Precision Liquids, Dufferin Road, Belfast Harbour Estate, Belfast
McIlmoyle, Dr W.A.	Animal Nutrition & Agric. Consultants, 2 Gregg Street, Lisburn, BT27 5AN
McLean, Mr B.	W L Duffield & Sons Ltd, Saxlingham Thorpe Mills, Norwich NR15 1TY
Miller, Mr C.	Waterford Foods plc, Main Street, Dungarvan, Co. Waterford, Ireland
Miller, Dr E.L.	Dept. Clinical Veterinary Medicine, Nutrition Laboratory, 307 Huntingdon Road, Cambridge CB3 0JQ
Mills, Mr C.	University of Nottingham, School of Agriculture, Sutton Bonington, Loughborough, Leics. LE12 5RD
Mitchell, Mr P.P.	Lopen Feed Mills Ltd, Mill Lane, Lopen, Somerset
Moore, Mr D.R.	David Moore (Flavours) Ltd, 29 High Street, Harpenden, Herts.AL5 2RU
Morgan, Dr J.T.	Four Gables, The Fossway, Stow on the Wold, Cheltenham, Gloucs GL54 1JU
Morris, Mr W.	Pauls Agriculture Ltd, P O Box 39, Key Street, Ipswich.
Moss, Mrs A.	ADAS Feed Evaluation Unit, Alcester Road, Stratford on Avon
Mounsey, Mr H.	HGM Publications, Abney House, Baslow, Derbyshire DE4 1RZ
Mounsey, Mr S.P.	HGM Publications, Abney House, Baslow, Derbyshire DE4 1RZ
Murray, Mr F.	Dairy Crest Ingredients, Philpot House, Rayleigh, Essex SS6 7HH
Newcombe, Mrs J.O.	University of Nottingham, School of Agriculture, Sutton Bonington, Loughborough, Leics LE12 5RD
Nicholson, Miss S.L.	BOCM Silcock, Basing View, Basingstoke, Hants RG21 2EQ
O'Grady, Dr J.	IAWS Group plc, 151 Thomas Street, Dublin 8, Ireland
Ogg, Miss A.	Harbro Farm Sales Ltd, Markethill, Turriff, Aberdeenshire

	AB53 7PA
Oldham, Dr J.D.	Scottish Agricultural College, Bush Estate, Penicuik, EH26 0QE
Orrell, Mr J.A.	The White Sea & Baltic Co Ltd, Arndale House, Otley Rd, Headingley, Leeds LS6 2UU
Orskov, Dr E.R.	Rowett Research Institute, Bucksburn, Aberdeen AB2 9SB
Overend, Dr M.A.	Nutec Limited, Eastern Avenue, Lichfield, Staffs
Owen, Mrs E.	M A F F, Room 303B, Ergon House, c/o Nobel House, 17 Smith Square, London SW1P 3JR
Owers, Dr M.	Pauls Agriculture Ltd, P O Box 39, Key Street, Ipswich
Packington, Mr A.	Colborn Dawes Nutrition Ltd, Heanorgate, Heanor, Derbys DE7 7SG
Papasolomontos, Dr S.	Dalgety Agriculture Ltd, 180 Aztec West, Almondsbury, Bristol BS12 4TH
Partridge, Dr G.G.	BP Nutrition (UK) Ltd, Wincham, Northwich, Cheshire CW9 6DF
Pass, Mr R.T.	Borthwick, 33 Ellersly Road, Edinburgh EH12 6JW
Pearce, Mr D.	Degussa Ltd, Paul Ungerer Ho, Wilmslow BK9 3RL
Pearson, Mr A.	Hoechst UK Ltd, Hoechst Animal Health, Walton Manor, Walton, Milton Keynes, Bucks MK7 7AJ
Perrott, Mr G.	Trident Foods, PO Box 11, Dundle Road, Peterborough PE2 9QU
Perry, Mr F.	BP Nutrition (UK) Ltd, Wincham, Northwich, Cheshire CW9 6DF
Pickford, Mr J.R.	Tecracon Ltd, Bocking Hall, Bocking Church Street, Braintree, Essex CM7 5JY
Pike, Dr I.H.	International Association of Fish Meal Manufacturers, Hoval House, Orchard Parade, Mutton Lane, Potters Bar, Herts EN6 3
Pine, Mr A.P.	Inst. of Ecology & Resource Management, University of Edinburgh, West Mains Road, Edinburgh EH9 3JG
Piva, Dr A.	Facolta di Agraria, Via E. Parmense, 84, 29100 Piacenza, Italy
Piva, Dr G.	Facolta di Agraria, Via E. Parmense, 84, 29100 Piacenza, Italy
Plowman, Mr G.B.	G W Plowman & Son Ltd, Selby House, High Street, Spalding, Lincs PE11 1TW
Powis, Miss J.	M A F F, Room 303B, Ergon House, c/o Nobel House, 17 Smith Square, London SW1P 3JR
Powles, Miss J.	University of Nottingham, Department of Agriculture and Horticulture, Sutton Bonington, Loughborough, Leics LE12 5RD
Putnam, Mr M.E.	Roche Products Ltd, P O Box 8, Welwyn Garden City, Herts AL7 3AY
Rae, Dr R.	Premier Nutrition Products Ltd, Unit 8, Walk Mill Business Park, Cannock, Staffs WS11 3XE
Raine, Dr H.	J Bibby Agriculture Ltd, Head Office, Adderbury, Oxon
Raper, Mr G.J.	Laboratories Pancosma (UK) Ltd, Crompton Road Industrial Estate, Ilkeston, Derbys DE7 4BG
Read, Mr M.	Smithkline Beecham Animal Health, Terrapin Building, Walton Oaks, Tadworth, Surrey KT20 7HE
Reeve, Mr J.G.	R.S. Feed Blocks, Orleigh Mill, Bideford, Devon
Retter, Dr W.C.	Heygate & Sons Ltd, Bugbrooke, Northampton
Rigg, Mr G.J.	Elanco Animal Health, Dextra Court, Chapel Hill, Basingstoke, Hants RG21 2SY
Robinson, Mr D.K.	Favor Parker Ltd, The Hall, Stoke Ferry, Kings Lynn,

	Norfolk PE33 9SE
Rosen, Dr G.D.	Field Investigations and Nutrition Services Ltd, 66 Bathgate Road, London SW19 5PH
Rosillo, Mr J.	University of Nottingham, School of Agriculture, Sutton Bonington, Loughborough, Leics LE12 5RD
Round, Mr J.	J Bibby Agriculture Ltd, Head Office, Adderbury, Oxon
Sanz, Mr J.M.	Calidad INVE Centro, B de la Estacion S/N, 45220-Yeles (Toledo), Spain
Scholman, Mr G.J.	Provimi BV, Veerlaan 17-23, 3072 An Rotterdam, The Netherlands
Scott, Mr M.R.	Fergusson Wild & Co Ltd, 3 St. Helen's Place, London EC3A 6BD
Serrano, Mr X.	Nutricio i Alimentacio Animal, Facultat Veterinaria-VAB, 08193 Bellaterra (Barcelona), Spain
Shipton, Mr P.	Dardis & Dunns Coarse Feeds Ltd, Ashbourne, Co. Meath, Eire
Shorrock, Dr C.	FSL Bells Ltd, Hartham Park, Corsham, Wilts
Shurlock, Dr T.	Smithkline Beecham Animal Health Ltd, Hunters Chase, Walton Oaks, Tadworth, Surrey KT20 7NT
Sinclair, Mr L.	University of Nottingham, School of Agriculture, Sutton Bonington, Loughborough, Leics LE12 5RD
Smith, Mr F.H.	Faculty of Veterinary Medicine, Ballsbridge, Dublin 4, Ireland
Smith, Mr M.P.	J Bibby Agriculture Ltd, Harlescott, Shrewsbury SY1 3TD
Sparks, Dr N.	Scottish Agricultural College, Poultry Science Department, Auchincruive, Ayr KA6 5HU
Speck, Mr P.	University of Nottingham, School of Agriculture, Sutton Bonington, Loughborough, Leics LE12 5RD
Spencer, Mr P.G.	Bernard Matthews plc, Gt Witchingham Hall, Norwich NR9 5QD
Stainsby, Mr A.K.	B A T A Ltd, Railway Street, Malton, N. Yorks YO17 0NU
Stark, Dr B.A.	Baydella, Bassetsbury Lane, High Wycombe, Bucks HP11 1RB
Statham, Mr R.	Pet's Choice, Greenbank Mill, Blackburn
Stebbens, Dr H.	Trident Feeds, P O Box 11, Oundle Road, Peterborough PE2 9QU
Stranks, Mr M.H.	ADAS, M.A.F.F., Burghill Road, Westbury-on-Trym, Bristol BS10 6NJ
Sutton, Dr J.D.	I G E R, Hurley, Maidenhead
Swarbrick, Mr J.E.	Borculo Whey Products UK Ltd, Bryman Four Estate, River Lane, Saltney, Chester
Tamminga, Dr S.	Wageningen Agricultural University, Dept. of Animal Nutrition, Haagsteeg 4, 6708 PM Wageningen, The Netherlands
Taylor, Dr A.J.	BOCM Silcock Ltd, Basing View, Basingstoke, Hants RG21 2EQ
Thelwall, Mr D.	Prospect Management Services, Prospect House, Copy Hewick, Ripon, N. Yorkshire HG4 5DB
Thomas, Dr C.	Scottish Agricultural College, Auchincruive, Ayr KA6 5HW
Thomas, Mr G.	Harvest Poultry, Green Farm, Kenninghall, Norfolk
Thompson, Mr D.	Rightfeeds Ltd, Woodlawn, Castlegarde, Cappamore, Co. Limerick, Ireland
Thompson, Dr F.	Rumenco (U M Group), Stretton House, Derby Rd, Burton on Trent, Staffs
Thompson, Mr J.	Feed Flavours (Europe) Ltd, Waterlip Mill, Cranmore,

	Shepton Mallet, Somerset BA4 4RN
Thompson, Mr R.J.	Amalgamated Farmers Ltd, Kinross, New Hall Lane, Preston, PR1 5JX
Thomson, Miss F.M.	BOCM Silcock Ltd, Silcock House, Basing View, Basingstoke, Hants RG21 2EQ
Thomson, Mr W.	Harbro Farm Sales Ltd, Markethill, Turriff, Abderdeenshire AB58 7PA
Thurston, Mr M.	Rhone Poulenc Chemicals, Oak House, Reeds Crescent, Watford WD1 1QH
Tolman, Mr H.	IVT-TNO/ILOB, P O Box 15, 6700 AA Wageningen, The Netherlands
Tonks, Mr W.P.	Park Tonks Ltd, 48 North Road, Gt. Abingdon, Cambridge
Toplis, Mr P.	SC Associates (Feedingstuffs) Ltd, The Limes, Sowerby Road, Sowerby, Thirsk, N Yorks YO7 1HX
Tucker, Miss S.A.	MAFF Gleadthorpe EHF, Meden Vale, Mansfield, Notts NG29 9PF
Twigge, Mr J.	BP Nutrition (UK) Ltd, Wincham, Northwich, Cheshire CW9 6DF
Uprichard, Mr J.	BP Nutrition (UK) Ltd, T/A BP Nutrition (Northern Ireland), 36 Ship Street, Belfast BT15 1JL
Van Den Broecke, Ir J.	Eurolysine, 16 Rue Ballu, 75009 Paris, France
Van der Ploeg, Ir H.	Trade Magazin de Molenaar, Stationsweg 4, 3603 EE, Maarsson, The Netherlands
Van Eys, Dr J.	Ralston Purina, 78056 St Quentin en Yuelines, Cedex, France
Vernon, Dr B.	Pauls Agriculture Ltd, P O Box 39, Key Street, Ipswich.
Verstegen, Dr M.W.A.	Wageningen Agricultural University, Haagsteeg 4, 6708 PM Wageningen, The Netherlands
Walker, Dr T.	Australian Poultry Ltd, Hawthrone Street, Beresfield, NSW,
Australia	
Wallace, Mr J.R.	Nutrition Trading (Int.) Ltd, Doric House, Church Street, Studley, Warwickshire B80 7LG
Ward, Mr J.	Frank Wright Ltd, Blenheim House, Blenheim Road, Ashbourne, Derbyshire DE6 1HA
Ward, Mr M.	Britphos Ltd, Rawdon House, Green Lane, Yeadon, Leeds LS19 7BY
Waterworth, Mr D.G.	ICI BioProducts, Billingham, Cleveland
Webster, Prof A.J.F.	University of Bristol, School of Veterinary Science, Langford, Bristol BS18 7DU
Webster, Mrs M.	Format International Ltd, Owen House, Heathside Crescent, Woking, Surrey GU22 7AG
Wilkinson, Dr J.I.D.	Lilly Research Centre Ltd, Erl Wood Manor, Windlesham, Surrey GU20 6PH
Wilkinson, Dr R.G.	Harper Adams Agricultural College, Newport, Shropshire TF10 8NB
Williams, Mr C.	Rhone-Poulenc Chemicals Ltd, ABM Brewing and Enzymes, Poleacre Lane, Woodley, Stockport, Cheshire SK6 1PQ
Williams, Mr D.J.	International Molasses, Shell Road, Royal Edward Dock, Avonmouth, Bristol BS11 9BW
Williams, Dr D.R.	BOCM Silcock Ltd, PO Box 4, Barlby Road, Selby, North Yorkshire YO8 7DT
Williams, Mr E.	J Bibby Agriculture Ltd, Sherburn in Elmet, Nr Leeds, Yorkshire
Williams, Mr J.S.	Four-F Nutrition, Darlington Road, Northallerton,

	N. Yorks DL6 2NW
Wilson, Mr B.	Nutec, Greenhills Centre, Greenhills Road, Tallaght, Dublin 24, Eire
Wilson, Dr B.J.	Cherry Valley Farms Ltd, N Kelsey Moor, Caistor, Lincs LN7 6HH
Wilson, Prof P.N.	E C R R, Crew Building, University of Edinburgh, West Mains Road, Edinburgh EH9 3JG
Winwood, Mr J.	Rhone-Poulenc Chemicals Ltd, ABM Brewing and Enzymes, Poleacre Lane, Wooley, Stockport, Cheshire SK6 1PQ
Wiseman, Dr J.	University of Nottingham, School of Agriculture, Sutton Bonington, Loughborough, Leics LE12 5RD
Woodford, Mr R.	Feed Flavours (Europe) Ltd, Waterlip Mill, Cranmore, Shepton Mallet, Somerset BA4 4RN
Woodward, Mr P.	Sun Valley Poultry Ltd, Sun Valley Feed Mill, Tram Inn, Allensmore, Hereford HR2 9AW
Woolford, Dr M.	Alltech UK, 16/17 Abenbury Way, Wrexham Ind. Estate, Wrexham, Clwyd LL13 9UZ
Zwart, Mr S.	Hydro Agri Rotterdam BV, Haassluissedijk 103, PO Box 58, 3130 AB Vlaardingen, The Netherlands

Index